T0190437

Streaming Linked Data

Riccardo Tommasini • Pieter Bonte •
Fabiano Spiga • Emanuele Della Valle

Streaming Linked Data

From Vision to Practice

 Springer

Riccardo Tommasini
LIRIS Lab
Institut National des Sciences Appliquées
Villeurbanne, France

Fabiano Spiga
Tallinn, Estonia

Pieter Bonte
Department of Information Technology
Ghent University
Ghent, Belgium

Emanuele Della Valle (iD)
Department of Electronics, Informatics and
Bioengineering
Politecnico di Milano
Milano, Italy

ISBN 978-3-031-15370-9   ISBN 978-3-031-15371-6   (eBook)
https://doi.org/10.1007/978-3-031-15371-6

This Springer imprint is published by the registered company Springer Nature Switzerland AG
The registered company address is: Gewerbestrasse 11, 6330 Cham, Switzerland

To my family, who always supported me in my journey, and to those who saw the potential in a rough stone and help polishing its talent. In particular, to Sherif Sakr (R.I.P.)

- Riccardo

To my loving wife and beautiful daughters, my sources of inspiration. Thank you for accompanying me along my path.

- Pieter

To my niece Sofia and to all the children of this world not suitable for children, who inspire men of good will to make it a better place.

- Fabiano

To my wife Simona and our wonderful daughter Giulia. Panta Rei, thus seize the day.

- Emanuele

Foreword

No matter what area of knowledge we may be talking about, it is always a challenging task to compile into a single book the most relevant knowledge accumulated during more than a decade of active research and application. Indeed, this books compiles and summarizes into a coherent storyline the work that was initiated more than 10 years ago by a group of researchers who had already embraced the Semantic Web principles and concepts to deal with the variety of data and identified the need to start addressing those aspects related to the velocity of data in many real-world contexts.

Having been part of this community since its early stages, including the supervision of PhD students and other early-stage researchers, and having collaborated with some of the authors of this book, as well as many others whose works are referenced in it, I must admit that I always found it difficult to collect in a single folder all the relevant reference materials that I should be providing to the next PhD student in this topic or to my students at the various research masters on Artificial Intelligence and Data Science that I participate in. Rooted on the work done for many years in tutorials around the topics of stream reasoning, RDF stream processing, and Streaming Linked Data, this book provides a clear solution for this problem and will become one of the key references in my courses around Semantic Web topics.

I want to emphasize here that this is not just a book that compiles disconnected pieces of research. By taking a quick look at the table of contents, the readers will immediately realize that the authors have gone far beyond generating an incremental compendium of the research works that have been done over the years, what would be relatively easy and would require little effort. They have created a coherent story where readers will first be able to understand the principles on which this research area has been established (around the aspects related to variety and velocity). Then they will be able to browse and learn from the historical evolution of approaches, well rooted on theory and practice. They will later understand what is the life cycle of streaming linked data projects and initiatives, based on the experience gathered by the authors in numerous projects. And finally they will understand which efforts have been done for the evaluation and continuous improvement of the theoretical frameworks and development systems around this area. All of this will provide those interested in getting up to speed into the world of streaming linked data to have a single source of reference.

And I am especially happy when seeing the latest chapter with the exercise book, something that in many situations we forget about when we are creating a book that may not only be used as a reference book for researchers but also as an initial textbook for practitioners. I will definitely use also the contents of the final chapter, the exercise book, with my Master students.

In summary, I recommend this book to all those interested in understanding how a semantic-based approach (and its corresponding technologies) can be applied in contexts that go beyond static data, either because they want to start applying these techniques in practice or because they are interested in pursuing research in the area. And I also recommend it for those like me who are teaching students at the Master level and want to provide them with an advanced textbook on a topic for which there was not such compiled material before.

Madrid, Spain   Oscar Corcho
June 2022   Ontology Engineering Group
                                    R&D Center for Artificial Intelligence (AI.nnovation Space)
                                                            Universidad Politécnica de Madrid

Preface

A fruitful sequence of tutorials prompted a rationalization of the research journey that, across 10 years, shepherded several researchers and led to numerous contributions. In 2008, when the Stream Reasoning research question was firstly envisioned, data velocity was starting to emerge. Highly dynamic resources did not yet populate the Web, but social networks and the Internet of Things pushed toward real-time data management. With the Large Knowledge Collider (LarKC) project, the vision became a community, which gained an identity within the Semantic Web context.

The second generation of researchers developed Stream Reasoning into RDF Stream Processing. Seminal contributions strengthen the theoretical foundation of the field, which expanded beyond the Semantic Web borders to mix with inductive and deductive artificial intelligence. Stream Reasoning can now count contributions in (Temporal) logic, Robotics, Machine Learning, and data integration.

RDF Stream Processing ultimately led to Streaming Linked Data, which this book is about. While the third generation of stream reasoning researchers is approaching seniority, best practices are emerging, and stream processing is rising as the industry's de facto standard for Big Data engineering and analytics.

This book was envisioned as a resource that collects research efforts and could guide the next generation of researchers, practitioners, and industrial stakeholders. Our goal is to explain the value of data integration when it does not neglect the time-varying nature of data and the continuous essence of some information needs.

This volume focuses on theoretical and practical aspects. It provides a comprehensive perspective of algorithms, systems, and applications for streaming linked data. Finally, it introduces the readers to RSP4J, a novel open-source project that aims to gather community efforts in software engineering and empirical research.

An introductory chapter positions the work by explaining what motivates the design of specific techniques for processing data streams using Web technologies. The prerequisite chapter briefly summarizes the necessary background concepts and models needed to understand the remaining content of the book. Subsequently, Chap. 3 focuses on processing RDF streams, taming data velocity in an open environment characterized by high data variety. It introduces query answering algorithms with RSP-QL and analytics functions over streaming data. Chapter 4 focuses on publishing streams and events on the Web as a prerequisite aspect to make data findable and accessible to applications. Chapter 5 touches on the problems of benchmarking systems that analyze Web streams to foster technological progress. It surveys existing benchmarks and introduces guidelines that may support new practitioners in approaching the issue of continuous analytics. Chapter 6 presents a list of examples and exercises that will help whoever approaches the area to get used to its practices and become confident in its technological stack.

This book provides a comprehensive overview of core concepts and technological foundations of continuous analytics of Web streams. It presents real-world examples and names systems and applications. Therefore, it is of particular interest to students, lecturers, and researchers in Web data management and stream data management.

In practice, this book would not have been possible without several people: (i) the authors, who are linked by more than just respect and professional appreciation; their research groups, namely, Marco Balduini, Daniele Dell'Aglio, Femke Ongenae, Alessandro Margara, Ahmed Awad, and Radwa ElShawi; and their collaborators within the stream reasoning community, who contributed to the vision scientifically, i.e., Jean-Paul Calbimonte, Danh LePhouc, Robin Kerskisaŕkka, Oscar Corcho, Alessandra Mileo, Ali Intizar, Jacopo Urbani, Boris Motik, Darko Anicic, Thomas Eiter, Patrik Schneider, and many many more.

The genesis of this book dates back to the end of 2019. A series of unfortunate events, including COVID-19 and the premature demise of Prof. Sherif Sakr,[1] slow down the writing process. Yet, we are glad to present this work to the community.

Last but not least, we thank our institutions, INSA Lyon and LIRIS Lab, Ghent University IMEC, Politecnico di Milano, and the University of Tartu. In particular, Riccardo Tommasini and Fabiano Spiga acknowledge support from the European Social Fund via IT Academy program. Pieter Bonte acknowledges his support from Research Foundation Flanders (FWO) through his postdoctoral fellowship (1266521N).

Lyon, France   Riccardo Tommasini
Gent, Belgium   Pieter Bonte
Tartu, Estonia   Fabiano Spiga
Milan, Italy   Emanuele Della Valle
April 2022

---

[1] Rest in Peace.

Contents

Acronyms

4Vs	Velocity, Volume, Variety, Veracity
ABox	Assertion Box
ASP	Answer Set Programming
BGP	Basic Graph Pattern 3
BigSPE	Big Stream Processing Engine
BN	Bayesian Network
CEP	Complex Event Processing
CPS	Cyber Physical System
CQA	Conjunctive Queries Answering
CQL	Continuous Query Language
CR	Continuous Reasoning
CSR	Cascading Stream Reasoning
DAG	Direct Acyclic Graph
DBMS	DataBase Management System
DCAT	Data Catalog Vocabulary
DL	Description Logic
DS	Design Science
DSMS	Data Stream Management System
ELF	Expressive Layered Firehose
EPL	Event Processing Language
ESR	Expressive Stream Reasoning
FCC	First Class Citizen
HMM	Hidden Markov Model
HTTP	Hyper-Text Transfer Protocol
ICM	Integrated Conceptual Model
IFP	Information Flow Processor
IoT	Internet of Things
IRI	Internationalized Resource Identifier
JSON-LD	JSON for Linked Data
KPIs	Key Performance Indicators
LD	Linked Data

LOD	Linked Open Data
LP	Linear Program
NoSR	Network of Stream Reasoners
N-Triples	N-Triples serialization format
OBERON	Ontology-Based Event RecognitiON
OWL	Web Ontology Language
PROV-O	PROV Ontology
QA	Query Answering
RA	Relational Algebra
RDD	Resilient Distributed Datasets
RDF	Resource Description Format
RDFS	RDF Schema
RSP	RDF Stream Processing
SCME	Single-Case Mechanism Experiments
SCRA	Systematic Comparative Research Approach
SDME	Statistical Difference-Making Experiments
SP	Stream Processing
SPARQL	SPARQL Protocol and RDF Query Language
SPE	Stream Processing Engine
SQL	Structured Query Language
SR	Stream Reasoning
SW	[Semantic Web]
TAR	Technical Action Research
TBox	Terminological Box
TP	Triple Pattern
TRA	Taxonomy Relational Algebra
TriG	TriG serialization format
Turtle	Terse RDF Triple Language
TVR	Time-varying Relation
URI	Uniform Resource Identifier
VoCaLS	Vocabulary for Cataloging Linked Streams
VoID	Vocabulary of Interlinked Datasets
VoIS	Vocabulary of Interlinked Streams
W3C	World Wide Web Consortium
WASP	Web Stream Processing Actor
WoD	Web of Data
WSP	Web Stream Processing
WWW	World Wide Web
YASPER	Yet Another Stream Processing Engine for RDF

Chapter 1
General Introduction

> *A process cannot be understood by stopping it. Understanding*
> *must move with the flow of the process, must join it and flow*
> *with it.*
>
> *—Frank Herbert, Dune*

Data never sleep.[1] It is impressive to see what happens in an Internet's minute. As shown in Fig. 1.1, for instance, from 2018 until 2021, almost 400,000 apps get each minute downloaded from the App Stores of Google and Apple. Seven million videos are viewed on YouTube. People worldwide watch 764,000 hours of Netflix. Furthermore, this is a growing trend. Between 2020 and 2021, there were 400,000 additional swipes per minute on Tinder and 800,000 additional views per minute on Twitch.

There are two ways of creating value from data streams as those illustrated above. On the one hand, analysts can record them for days, months, and years before processing them. This kind of approach is called historical analysis. In Big Data terminology [31], it relates to the *Volume* dimension, i.e., the need to work with data at scale (e.g., petabytes or exabyte of data). Best of breed Map Reduce implementations such as Apache Spark [48] and Apache Flink [13] can tame the Volume dimension. On the other hand, analysts may want to process this data as they flow. With a latency of milliseconds, seconds, or minutes, we can analyze them to trigger reactive decisions, to act in time. In Big Data terminology [31], it relates to the *Velocity* dimension, i.e., the need to work with data before they are no longer valuable. Stream and complex event processing technologies can tame the Velocity dimension [14]. Frameworks such as KSQL DB [25] and Structured Streaming [3] even allow for taming the volume and velocity dimensions simultaneously.

Unfortunately, addressing Velocity and Volume simultaneously is a matter of trade-offs. Figure 1.2a illustrates an important constraint: the faster we need to process data, the smaller the data are that we can process. Of course, such trade-off keeps moving toward faster processing and larger amounts of data, but even today

---

[1] https://www.visualcapitalist.com/what-happens-in-an-internet-minute-in-2019/.

Fig. 1.1 Internet in a minute: from 2018 to 2021[1]

Fig. 1.2 Taming Volume, Velocity, and Variety at the same time. (**a**) The trade-off between Volume and Velocity. (**b**) Variety constrains the trade-off, too

it is impossible to perform even embarrassingly parallel tasks as filtering flows of exabytes per seconds of data in a dedicated cluster.

In practice, large volumes of data typically show higher variety, because they arrive from different sources. In Big Data terminology [31], it relates to the Variety dimension, i.e., the need for integrating structured (e.g., relational), semi-structured (e.g., XML and JSON documents), and unstructured data formats (e.g., texts, speeches, images, and videos). Semantic Web technologies [2] can tame the Variety dimension. However, as Fig. 1.2b shows, they also bring in their own constraints: the higher the Variety of data to process, the stricter the trade-off between Volume and Velocity.

Still there are use cases that demand to address the three Vs simultaneously. In these scenarios, a need is emerging for analyzing a variety of fast flowing data at scale as soon as possible because they rapidly lose **value**. Siemens recently wrote about an industrial scenario where several pieces of industrial equipment such as turbines need continuous analytics [29]. Other authors reported similar needs in smart cities [8, 23, 36], social media analytics [5, 6], and power distribution [9]. Table 1.1 illustrates the typical requirements that arose when all Vs occur simultaneously.

Table 1.1 Requirements of an analytics solution that needs to tame Volume (Vol., in the table), Velocity (Vel.), and Variety (Var.) simultaneously

Requirement	Example	Vol.	Vel.	Var.
Massive data	Typical IoT platform is connected to hundreds of thousands of IoT sensors; a regular social media platform has billions of monthly active users, etc.	✓		
Streaming data	A typical IoT platform may receive tens of thousands of events per second; millions of users interact with a standard social media platform every minute; etc.		✓	
Heterogeneous data	A large variety of static and streaming data sources and data management solutions exist in any domain. Any analysis nowadays is cross-data source, if not cross-data platform			✓
Incomplete data	IoT sensors can run out of power, or network links can break down; luckily enough, not all the conversations happen exclusively on social media, etc.			✓
Reactive answers	Analytics solutions must generate answers while meeting operational deadlines set by event occurrence. The time available to answer depends on the application domain: routing decision in a contact center should be sub-second; in predictive maintenance, the detection of faults must occur within tens of minutes		✓	✓
Fine-grained information access	The analysis may not only operate at high level of aggregation, but a query may require to locate exactly a piece of equipment or a user among tens of thousands of similar ones	✓	✓	✓
Complex domain models	Social media analytics may require topic models to make sense of a conversation, and an IoT analytics solution may require to model operational and control processes, etc.			✓

? Stream Reasoning Research Question

Is it possible to make sense in real time of multiple, heterogeneous, gigantic, and inevitably noisy and incomplete data streams in order to support the decision process of extremely large numbers of concurrent users?

Stream Reasoning [44] is the field that studies solutions that simultaneously address those requirements. The Stream Reasoning concept originated in 2008 when E. Della Valle, S. Ceri, F. van Harmelen, and D. Fensel elicited the requirements above and assessed the ability of existing technologies in addressing them (see the first two columns of Table 1.2). The lack of a solution able to address them all encouraged E. Della Valle et al. to formulate the Stream Reasoning's research question.

At first glance, answering such research question may appear impracticable due to the conflicting requirements. However, at a closer look, it does not differ much from other problems in Computer Science that requires to find solutions that trade-off between two or more conflicting requirements. Take, for instance, the

Table 1.2 Requirements coverage: Stream Processing vs. Semantic Web vs. SR

Requirement	Stream processing	Semantic Web	Stream reasoning
Massive data	✓	✓	✓
Streaming data	✓	✗	✓
Heterogenous data	✗	✓	✓
Incomplete data	✗	✓	✓
Reactive answers	✓	✗	✓
Fine-grain information access	✓	✓	✓
Complex domain models	✗	✓	✓

memory access problem. While the general problem of accessing with low latency an arbitrary large amount of data is not solvable, memory hierarchies (see [43] p. 30) offer a practical solution. They pose restrictions on the size and capabilities of each level, but at the same time they guarantee performances to algorithms designed to exploit data locality.

1.1 Web Stream Processing at Glance

Without anticipating too many concepts from the core of the book, this section aims at convincing the reader that, after almost 15 years of joint efforts between the AI and the Stream Processing communities, the Stream Reasoning research question was broadly investigated [20, 35]. In practice, Web Stream Processing (WSP) technologies, which this book presents, are the most tangible results of the Stream Reasoning research.

WSP positions itself within the Semantic Web research context. It encompasses continuous extensions of both the Semantic Web data model RDF, namely, RDF Streams, and its query language SPARQL, in particular RSP-QL. As shown in Fig. 1.3, RDF Streams and RSP-QL [19] are, by far, the most prominent attempts toward an industrial standardization of Stream Reasoning concepts.

At the time of writing, Web Stream Processing has been used to solve a variety of use cases, typically solving a data integration problem on top of one or more data streams, optionally in combination with slower changing data. It has been used for *monitoring wind turbines* [28, 29], to identify turbines whose sensor readings are similar to other turbines that subsequently had critical failures. In *oil rig drilling* [27], it was used to determine how long, according to historical records, one can keep drilling when the sensors are indicating that it is about to get stuck. *Smart Cities* [1, 11, 34, 36] are a domain that is ideal for Web Stream Processing due to the various data sources that need to be integrated. In *Smart Cities*, it has been deployed to solve questions such as what are the best routes to commute home, without running into traffic jams, given the current weather and traffic conditions or where would be the best location to spend my evening given the presence of

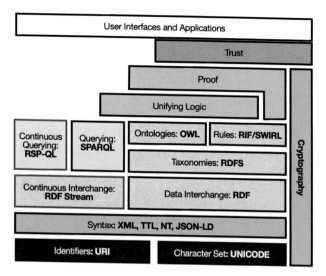

Fig. 1.3 Extended Semantic Web stack

people and their activities. *Social Media* [7, 26] is another popular domain in Web Stream Processing applications. Questions that are being solved using it consist of finding who the top influencers are that drive the discussions regarding the current top emerging topics across different social networks. Web Stream Processing has been used in various *Health* [10, 16, 17, 21] use cases, investigation questions such as are the vitals that are being monitored of each person in line with what can be expected given the specific pathology of each patient or which caregiver is best suited (based on their location, schedule and capabilities) to check up on a specific patient? Web Stream Processing stream processing has also been used for real-time monitoring and training adaptation of cyclists [15]. Last but not least, in *marine traffic monitoring* [18], it has been used for tracking purposes of vessels and reactive incident handling.

1.2 The ColorWave Running Example

This section introduces a running example that will help explaining concepts related to Stream Processing, Semantic Web, and Web Stream Processing. The example is agnostic from any existing application domain, since in the authors' experience, it helps focusing on the specific aspects of streaming data management, without introducing many details that may distract the readers and hinder their comprehension.

In practice, the example consists of an application processing a stream of colors (or colored boxes as depicted in Fig. 1.4). Each element is a timestamped color

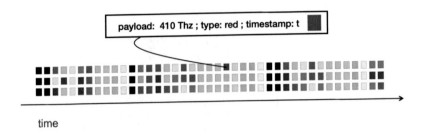

time

Fig. 1.4 An stream of colored boxes

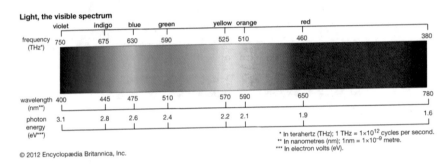

Fig. 1.5 Light spectrum overview

observation made by a sensor (e.g., red at time t). Nevertheless, to make the example more realistic, it is possible to present some contextual information. For instance, it is reasonable to imagine a device capable of sensing the light spectrum. Figure 1.5 helps mapping the sensor measurement into symbolic representations of colors, e.g.:

- if the light frequency is between 650 and 605, the light is perceived as *blue*;
- if the light frequency is between 605 and 545, the light is perceived as *green*;
- if the light frequency is between 545 and 480, the light is perceived as *yellow*;
- if the light frequency is between 480 and 380, the light is perceived as *red*.

The next chapters explain that stream processing is about analyzing data as soon as they arrive. In the ColorWave example, it translates into asking questions like in Fig. 1.6, e.g., "how many red color observations were there in the last minute?".

Without exacerbating the example complexity, it is possible to imagine an application that needs to interpret the sensor measurements. In particular, two alternative interpretations of the colors are relevant. Figure 1.7 describes the colors' structure, i.e., their temperature as warm and cold, and their composition, i.e., primary colors can be combined to define secondary, tertiary, and even n-ary combinations. Similarly, Fig. 1.8 describes the relation between colors and sentiment, e.g., it associates red with anger and green with happiness. According to such additional information, it is possible to refine the analyses in Fig. 1.7 as in

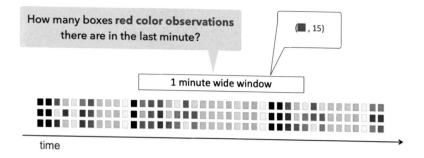

Fig. 1.6 Counting red occurrences in the stream of colors

Fig. 1.7 The structure of colors

Fig. 1.8 The sentiment of colors

Fig. 1.9, which shows how to combine the streaming data from the sensors with domain knowledge.

1.3 RDF Stream Processing with RSP4J

This book aims at providing both the theoretical and the practical knowledge required to develop Web Stream Processing application. In particular, for the latter, the book builds around the RSP-QL reference model from Dell'Aglio et al. [19],

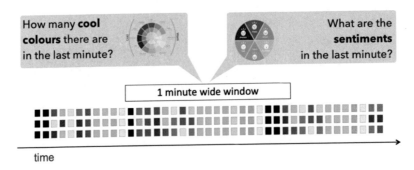

Fig. 1.9 Information needs for the color example

which aims at unifying prior works on the matter, and RSP4J, i.e., a an API for the development of RSP applications and engines [42]. As a comparison, the RSP4J is similarly to the OWL API [24], which helped providing a unified entry point in the context of OWL reasoning. The remainder of the section introduces the library.

1.3.1   Challenges and Requirements

Over the years, a variety of RSP engines has been proposed, e.g., C-SPRQL engine [4], CQELS [32], and Strider [37]. However, developing these prototypes unveiled a number of challenges that later foster the design of RSP4J. Such challenges are presented below, dividing them into three main categories.

Fast Prototyping In 2008, the first Stream Reasoner prototype came out [47]. Since then, the SR community has designed a number of working prototypes [4, 12, 32, 37], with the intent of proving the feasibility of the vision. E-health, smart cities, and financial transaction are examples of use cases where such prototypes were successfully used. Despite many attempts, the efforts needed in designing and engineering good prototypes are still extremely high; moreover, their maintenance is often unsustainable for researchers. As a matter of fact, prototypes are often designed with a minimal set of requirements and without shared design principles. In such scenarios, (C1) adding new operators, (C2) new data sources to consume, or (C3) experimenting with new optimizations techniques either requires huge manual efforts or is almost impossible.

Comparable Research & Benchmarking Apart from developing proofs of concept, the SR/RSP communities have focused a lot on Comparative Research (CR) [38, 39] and benchmarking [1, 30, 33, 40, 49]. CR studies the differences and similarities across SR/RSP approaches. Stream Reasoners and RSP engines can only be compared when they employ the same semantics. Thus, a fair comparison demands a deep theoretical comprehension of the approaches, a proper formulation

Table 1.3 Challenges v. requirements

Challenges →	Fast prototyping			Comparable research				Dissemination		
Requirements↓	C1	C2	C3	C4	C5	C6	C7	C8	C9	C10
R_1 Extensible architecture	✓	✓	✓		✓		✓			
R_2 Declarative access				✓				✓		✓
R_3 Programming abstractions								✓	✓	✓
R_4 Experimentation			✓	✓	✓					
R_5 Observability	✓	✓	✓			✓	✓			

of the task to solve, and an adequate experimentation environment [41]. RSP-QL provides the needed theoretical foundation for this. Consequently, it is currently hard to (C4) reproduce the behavior of existing approaches in a comparable way. Moreover, experimentation is limited by (C5) the lack of parametric solutions, i.e., the configurability of the operators allowing to match engine behavior. On the other hand, benchmarking aims at pushing the technological progress by guaranteeing a fair assessment. While some of the challenges are shared with CR [39, 41], benchmarking is empirical research. To this extent, monitoring both (C6) the execution of continuous queries and (C7) the engine behavior at run time is of paramount importance. Unfortunately, not all the existing prototypes provide such entry points, and only black box analysis is possible, e.g., it is impossible to measure the performance of each of the engine's internal operators.

Dissemination Although SR research is at its infancy, a lot has been done on the teaching side. As prototypes and approaches reach maturity, several tutorials and lectures were delivered at major venues, including ICWE, ESWC, ISWC, RW, and TheWebConf [22, 45, 46]. These tutorials were often practical and aimed at engaging with their audience using simple yet meaningful applications. Nevertheless, existing prototypes were not designed for teaching purposes. Thus, they lack important features like the possibility to (C8) inspect the engine behaviors, and (C9) they are not designed to ease the understanding at various levels of abstraction. Indeed, prototypes often (C10) neglect their full compliance to the underlying theoretical framework for practical reasons. Although this approach often benefits performance, it makes for a steeper learning curve.

Table 1.3 contrasts the challenges just presented with the requirements for an API to satisfy. Although they could be generalized for any RSP engine and Stream Reasoner, the focus is restricted to Window-based RDF Stream Processing Engines, i.e., those covered by the RSP-QL specification.

R_1 **Extensible Architecture**. An RSP API should allow the easy addition of data sources (C2) and operators (C1) and the design of optimization techniques (C3). Moreover, an RSP API should grant experimentation by allowing the addition of execution parameters (C5) and ease the extension of engine capabilities (C7) making the streams' abstraction accessible and extendable allows the implementation of a variety of streams, ranging from Kafka streams to Web sockets.

R$_2$ **Declarative Access**. An RSP API should be accessible in a declarative and configurable manner (C4). It should allow querying according to a formal semantics, e.g., RSP-QL (C10), and should allow controlling the engine and the query life cycles (C8) the Querying component allows the definition of tasks in a declarative manner.

R$_3$ **Programming Abstractions**. An RSP API should provide programming abstractions that allow interacting with the engine at various levels of abstractions (C9), abstractions that are based on a theoretical framework (C10), and that provide a blueprint to make sense of the engine behavior (C8)

R$_4$ **Experimentation**. An RSP API should be suitable for experimentation; thus, it should foster comparative research. To this extent, it should allow experimentation with optimizations techniques (C3), enabling to execute experiments using alternative configurations (C5). Last but not least, the reproducibility of state-of-the-art solutions should be a priority so to enable replication studies (C4)

R$_5$ **Observability**. An RSP API should be observable by design, enabling the collection of metrics at different levels, i.e., stream level, operator level, query level (C6), and engine level (C7). Observability should be independent from architectural changes (C1 and C2) and ease studying for optimizations (C3).

1.3.2   RSP4J's Architecture

The above extracted requirements are at the heart of RSP4J. Figure 1.10 depicts RSP4J's architecture. Each model tackles a number of the above requirements:

- **Querying:** provides the functionality to enable declarative access (R$_2$).

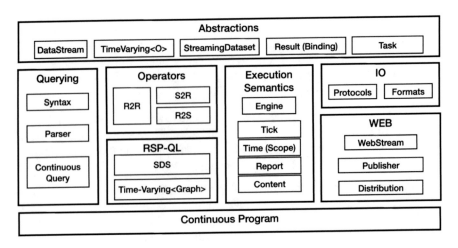

Fig. 1.10 RSP4J's architecture

- **RSP-QL:** models additional RSP-QL abstractions (R_3) such as the Streaming Dataset (SDS) and the Time-varying graph.
- **Operators:** RSP4J provides three operator abstractions, *StreamToRelation*, *RelationToRelation*, and *RelationToStream*, in line with their theoretical foundations (R_3) and provides these operators as an entry point for extensions and optimizations (R_1), which could be monitored independently (R_5).
- **Execution Semantics:** this module includes the abstractions to control and monitor the engine and the query life cycle (R_4 and R_5).
- **IO:** contains all the functionality to consume streams from various sources and eases the integration of new sources (R_1).
- **Web:** provides all the means for publishing streams on the Web (R_3).
- **Abstractions:** includes the abstract concepts reused throughout the API (R_3).
- **Continuous Program:** represents the ever-lasting computation required by continuous queries. It allows monitoring and controlling the query life cycle (R_5.

Throughout this book, RSP4J will be used as a practical counterpart to the theoretical foundations. As RSP4J has been built specifically for dissemination purposes, it is ideal for this book where we will peek under the hood of RSP engines.

1.4 How to Read This Book

Each chapter of this book can be read independently; however, the authors suggest to first go through the Preliminaries (Chap. 2) in order to master all the background that is needed to understand the remainder of the book.

Throughout the book, certain paragraphs are highlighted. The meaning of each highlighted paragraph can be interpreted as follows:

> **? Questions**

Warning boxes like this are used to attract the readers attention on difficult passages or details.

> **! Warnings**

These paragraphs are used to catch the reader's attention in order to warn them regarding a certain topic.

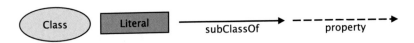

Fig. 1.11 Legend for the ontology schemas

> **Exercises**

These paragraphs highlight exercises that the reader can conduct in order to practice the learned material.

Hands-on Practice

These paragraphs contain a number of hands-on examples that the reader can follow along. Most of these hands-on examples will be supported by RSP4J.

Tips

These paragraphs highlight a suggestion from the authors.

Throughout the book, same structure is adhered when introducing ontology schemas. Figure 1.11 shows the different arrows, either denoting subclass of relation (with full arrows) or ontology properties (with dotted arrows). Note that the latter can be both object and data properties. A distinction is made between classes (ovals) and literal values (with rectangles).

Furthermore, a number of prefixes are used throughout the book, and a complete list can be found below as a reference:

- rdf:http://www.w3.org/1999/02/22-rdf-syntax-ns#
- rdfs: http://www.w3.org/2000/01/rdf-schema#
- owl: http://www.w3.org/2002/07/owl#
- xsd: http://www.w3.org/2000/10/XMLSchema#
- ssn: http://www.w3.org/ns/ssn/
- ces: http://www.insight-centre.org/ces#
- sao: http://purl.oclc.org/NET/sao#
- ct:http://www.insight-centre.org/citytraffic#
- frappe: http://streamreasoning.org/ontologies/frappe#
- event: http://purl.org/NET/c4dm/event.owl#
- time: http://www.w3.org/2006/time#
- tl: http://purl.org/NET/c4dm/timeline.owl#
- geo: http://www.w3.org/2003/01/geo/wgs84_pos#
- prov: http://www.w3.org/ns/prov#
- vocals: http://w3id.org/rsp/vocals#

1.5 Outline

The remainder of the book is organized as follows:

- Chapter 2 presents the background needed to understand the remainder of the book.
- Chapter 3 focuses on how *Variety* and *Velocity* can be tamed simultaneously.
- Chapter 4 introduces a life cycle for Streaming Linked Data.
- Chapter 5 presents existing Web Stream Processing Systems and Benchmarks.
- Chapter 6 details a number of hands-on exercises.

References

1. Ali, Muhammad Intizar, Feng Gao, and Alessandra Mileo. 2015. Citybench: A configurable benchmark to evaluate RSP engines using smart city datasets. In *The Semantic Web - ISWC 2015 - 14th International Semantic Web Conference, Bethlehem, PA, USA, October 11–15, 2015, Proceedings, Part II*, 374–389.
2. Antoniou, Grigoris, Paul Groth, Frank van Harmelen, and Rinke Hoekstra. 2012. *A Semantic Web Primer*. 3rd ed. Cambridge: MIT Press.
3. Armbrust, Michael, Tathagata Das, Joseph Torres, Burak Yavuz, Shixiong Zhu, Reynold Xin, Ali Ghodsi, Ion Stoica, and Matei Zaharia. 2018. Structured streaming: A declarative API for real-time applications in apache spark. In *SIGMOD Conference*, 601–613. New York: ACM.
4. Barbieri, Davide Francesco, Daniele Braga, Stefano Ceri, Emanuele Della Valle, and Michael Grossniklaus. 2010. C-SPARQL: A continuous query language for RDF data streams. *International Journal of Semantic Computing* 4 (1): 3–25.
5. Barbieri, Davide Francesco, Daniele Braga, Stefano Ceri, Emanuele Della Valle, Yi Huang, Volker Tresp, Achim Rettinger, and Hendrik Wermser. 2010. Deductive and inductive stream reasoning for semantic social media analytics. *IEEE Intelligent Systems* 25 (6): 32–41.
6. Balduini, Marco, Irene Celino, Daniele Dell'Aglio, Emanuele Della Valle, Yi Huang, Tony Lee, Seon-Ho Kim, and Volker Tresp. 2012. Bottari: An augmented reality mobile application to deliver personalized and location-based recommendations by continuous analysis of social media streams. *Journal of Web Semantics* 16: 33–41.
7. Balduini, Marco, Irene Celino, Daniele Dell'Aglio, Emanuele Della Valle, Yi Huang, Tony Lee, Seon-Ho Kim, and Volker Tresp. 2014. Reality mining on micropost streams. *Semantic Web* 5 (5): 341–356.
8. Balduini, Marco, Marco Brambilla, Emanuele Della Valle, Christian Marazzi, Tahereh Arabghalizi, Behnam Rahdari, and Michele Vescovi. 2019. Models and practices in urban data science at scale. *Big Data Research* 17: 66–84.
9. Belcao, Matteo, Emanuele Falzone, Enea Bionda, and Emanuele Della Valle. Chimera: A bridge between big data analytics and semantic technologies. In *ISWC*. Vol. 12922. *Lecture Notes in Computer Science*, 463–479. Berlin: Springer.
10. Bonte, Pieter, Femke Ongenae, and Filip De Turck. 2019. Subset reasoning for event-based systems. *IEEE Access* 7: 107533–107549.
11. Bonte, Pieter, Mathias De Brouwer, Dörthe Arndt, Miel Vander Sande, Pieter Heyvaert, Anastasia Dimou, Pieter Colpaert, Ruben Verborgh, Filip De Turck, and Femke Ongenae. 2020. Context-aware route planning: a personalized and situation-aware multi-modal transport routing approach. In *ISWC2020, the International Semantic Web Conference*, 1–5.

12. Calbimonte, Jean-Paul, Hoyoung Jeung, Oscar Corcho, and Karl Aberer. 2012. Enabling query technologies for the semantic sensor web. *International Journal On Semantic Web and Information Systems (IJSWIS)* 8 (1): 43–63.

13. Carbone, Paris, Asterios Katsifodimos, Stephan Ewen, Volker Markl, Seif Haridi, and Kostas Tzoumas. 2015. Apache flink™: Stream and batch processing in a single engine. *Bulletin of the IEEE Computer Society Technical Committee on Data Engineering* 38 (4): 28–38.

14. Cugola, Gianpaolo, and Alessandro Margara. 2012. Processing flows of information: From data stream to complex event processing. *ACM Computing Surveys* 44 (3): 15:1–15:62.

15. Daneels, Glenn, Esteban Municio, Kathleen Spaey, Gilles Vandewiele, Alexander Dejonghe, Femke Ongenae, Steven Latré, and Jeroen Famaey. 2017. Real-time data dissemination and analytics platform for challenging iot environments. In *2017 Global Information Infrastructure and Networking Symposium (GIIS)*, 23–30. Piscataway: IEEE.

16. De Brouwer, Mathias, Femke Ongenae, Pieter Bonte, and Filip De Turck. 2018. Towards a cascading reasoning framework to support responsive ambient-intelligent healthcare interventions. *Sensors* 18 (10): 3514.

17. De Brouwer, Mathias, Pieter Bonte, Dörthe Arndt, Miel Vander Sande, Pieter Heyvaert, Anastasia Dimou, Ruben Verborgh, Filip De Turck, and Femke Ongenae. 2020. Distributed continuous home care provisioning through personalized monitoring & treatment planning. In *Companion Proceedings of the Web Conference 2020*, 143–147.

18. Dejonghe, Alexander, Femke Ongenae, Stijn Verstichel, and Filip De Turck. 2017. C-GeoSPARQL: Streaming geosparql support on C-SPARQL. In *Joint Proceedings of the 2nd RDF Stream Processing (RSP 2017) and the Querying the Web of Data (QuWeDa 2017) Workshops co-located with 14th ESWC 2017 (ESWC 2017)*, 1–10.

19. Dell'Aglio, Daniele, Emanuele Della Valle, Jean-Paul Calbimonte, and Óscar Corcho. 2014. RSP-QL semantics: A unifying query model to explain heterogeneity of RDF stream processing systems. *International Journal on Semantic Web and Information Systems (IJSWIS)* 10 (4): 17–44.

20. Dell'Aglio, Daniele, Emanuele Della Valle, Frank van Harmelen, and Abraham Bernstein. 2017. Stream reasoning: A survey and outlook. *Data Science* 1 (1–2): 59–83.

21. Dobbins, Chelsea, Paul Fergus, Madjid Merabti, and David Llewellyn-Jones. 2012. Monitoring and measuring sedentary behaviour with the aid of human digital memories. In *2012 IEEE Consumer Communications and Networking Conference (CCNC)*, 395–398. Piscataway: IEEE.

22. Falzone, Emanuele, Riccardo Tommasini, and Emanuele Della Valle. 2020. Stream reasoning: From theory to practice. In *Reasoning Web International Summer School*. Vol. 12258, 85–108. Berlin: Springer.

23. Gao, Feng, Muhammad Intizar Ali, and Alessandra Mileo. 2014. Semantic discovery and integration of urban data streams. In *Proceedings of the Fifth Workshop on Semantics for Smarter Cities a Workshop at the 13th International Semantic Web Conference (ISWC 2014), Riva del Garda, Italy, October 19, 2014*, 15–30.

24. Horridge, Matthew, and Sean Bechhofer. 2011. The OWL API: A java API for OWL ontologies. *Semantic Web* 2 (1): 11–21.

25. Jafarpour, Hojjat, and Rohan Desai. 2019. KSQL: streaming SQL engine for apache kafka. In *Advances in Database Technology - 22nd International Conference on Extending Database Technology, EDBT 2019, Lisbon, Portugal, March 26–29, 2019*, ed. Melanie Herschel, Helena Galhardas, Berthold Reinwald, Irini Fundulaki, Carsten Binnig, and Zoi Kaoudi, 524–533. OpenProceedings.org.

26. Keskisärkkä, Robin, and Eva Blomqvist. 2013. Semantic complex event processing for social media monitoring-a survey. In *Proceedings of Social Media and Linked Data for Emergency Response (SMILE) Co-located with the 10th Extended Semantic Web Conference, Montpellier, France. CEUR Workshop Proceedings (May 2013)*.

27. Kharlamov, Evgeny, Dag Hovland, Ernesto Jiménez-Ruiz, Davide Lanti, Hallstein Lie, Christoph Pinkel, Martin Rezk, Martin G Skjæveland, Evgenij Thorstensen, Guohui Xiao, et al. 2015. Ontology based access to exploration data at statoil. In *International Semantic Web Conference*, 93–112. Berlin: Springer.

28. Kharlamov, Evgeny, Yannis Kotidis, Theofilos Mailis, Christian Neuenstadt, Charalampos Nikolaou, Özgür Özcep, Christoforos Svingos, Dmitriy Zheleznyakov, Sebastian Brandt, Ian Horrocks, et al. 2016. Towards analytics aware ontology based access to static and streaming data. In *International Semantic Web Conference*, 344–362. Berlin: Springer.

29. Kharlamov, Evgeny, Yannis Kotidis, Theofilos Mailis, Christian Neuenstadt, Charalampos Nikolaou, Özgür L. Özçep, Christoforos Svingos, Dmitriy Zheleznyakov, Yannis E. Ioannidis, Steffen Lamparter, Ralf Möller, and Arild Waaler. 2019. An ontology-mediated analytics-aware approach to support monitoring and diagnostics of static and streaming data. *Journal of Web Semantics* 56: 30–55.

30. Kolchin, Maxim, Peter Wetz, Elmar Kiesling, and A. Min Tjoa. 2016. Yabench: A comprehensive framework for RDF stream processor correctness and performance assessment. In *Web Engineering - 16th International Conference, ICWE 2016, Lugano, Switzerland, June 6–9, 2016. Proceedings*, 280–298.

31. Laney, Doug. 2001. 3d data management: Controlling data volume, velocity and variety. *META Group Research Note* 6 (70): 1.

32. Le-Phuoc, Danh, Minh Dao-Tran, Josiane Xavier Parreira, and Manfred Hauswirth. 2011. A native and adaptive approach for unified processing of linked streams and linked data. In *International Semantic Web Conference*, 370–388. Berlin: Springer.

33. Le Phuoc, Danh, Minh Dao-Tran, Minh-Duc Pham, Peter A. Boncz, Thomas Eiter, and Michael Fink. 2012. Linked stream data processing engines: Facts and figures. In *International Semantic Web Conference (2)*. Vol. 7650. *Lecture Notes in Computer Science*, 300–312. Berlin: Springer.

34. Lécué, Freddy, Simone Tallevi-Diotallevi, Jer Hayes, Robert Tucker, Veli Bicer, Marco Sbodio, and Pierpaolo Tommasi. 2014. Smart traffic analytics in the semantic web with star-city: Scenarios, system and lessons learned in dublin city. *Journal of Web Semantics* 27: 26–33.

35. Margara, Alessandro, Jacopo Urbani, Frank van Harmelen, and Henri E. Bal. 2014. Streaming the web: Reasoning over dynamic data. *Journal of Web Semantics* 25: 24–44.

36. Puiu, Dan, Payam M. Barnaghi, Ralf Toenjes, Daniel Kuemper, Muhammad Intizar Ali, Alessandra Mileo, Josiane Xavier Parreira, Marten Fischer, Sefki Kolozali, Nazli FarajiDavar, Feng Gao, Thorben Iggena, Thu-Le Pham, Cosmin-Septimiu Nechifor, Daniel Puschmann, and João Fernandes. 2016. Citypulse: Large scale data analytics framework for smart cities. *IEEE Access* 4: 1086–1108.

37. Ren, Xiangnan, and Olivier Curé. 2017. Strider: A hybrid adaptive distributed rdf stream processing engine. In *International Semantic Web Conference*, 559–576. Berlin: Springer.

38. Scharrenbach, Thomas, Jacopo Urbani, Alessandro Margara, Emanuele Della Valle, and Abraham Bernstein. 2013. Seven commandments for benchmarking semantic flow processing systems. In *The Semantic Web: Semantics and Big Data, 10th International Conference, ESWC 2013, Montpellier, France, May 26-30, 2013. Proceedings*, 305–319.

39. Tommasini, Riccardo, Emanuele Della Valle, Marco Balduini, and Daniele Dell'Aglio. 2016. Heaven: A framework for systematic comparative research approach for RSP engines. In *ESWC*. Vol. 9678. *Lecture Notes in Computer Science*, 250–265. Berlin: Springer.

40. Tommasini, Riccardo, Marco Balduini, and Emanuele Della Valle. 2017. Towards a benchmark for expressive stream reasoning. In *RSP+QuWeDa@ESWC*. Vol. 1870. *CEUR Workshop Proceedings*, 26–36. CEUR-WS.org.

41. Tommasini, Riccardo, Emanuele Della Valle, Andrea Mauri, and Marco Brambilla. 2017. RSPLab: RDF stream processing benchmarking made easy. In *ISWC*, 202–209.

42. Tommasini, Riccardo, Pieter Bonte, Femke Ongenae, and Emanuele Della Valle. 2021. RSP4J: an API for RDF stream processing. In *The Semantic Web - 18th International Conference, ESWC 2021, Virtual Event, June 6–10, 2021, Proceedings*ed. Ruben Verborgh, Katja Hose, Heiko Paulheim, Pierre-Antoine Champin, Maria Maleshkova, Óscar Corcho, Petar Ristoski, and Mehwish Alam. Vol. 12731. *Lecture Notes in Computer Science*, 565–581. Berlin: Springer.

43. Toy, Wing N., and Benjamin Zee. 1986. *Computer Hardware-Software Architecture*. Hoboken: Prentice Hall Professional Technical Reference.

44. Valle, Emanuele Della, Stefano Ceri, Frank van Harmelen, and Dieter Fensel. 2009. It's a streaming world! reasoning upon rapidly changing information. *IEEE Intelligent Systems* 24 (6): 83–89.

45. Valle, Emanuele Della, Daniele Dell'Aglio, and Alessandro Margara. 2016. Taming velocity and variety simultaneously in big data with stream reasoning: Tutorial. In *DEBS*, 394–401. New York: ACM.

46. Valle, Emanuele Della, Riccardo Tommasini, and Marco Balduini. 2018. Engineering of web stream processing applications. In *Reasoning Web*. Vol. 11078. *Lecture Notes in Computer Science*, 223–226. Berlin: Springer.

47. Walavalkar, Onkar, Anupam Joshi, Tim Finin, Yelena Yesha, et al. 2008. Streaming knowledge bases. In *Proceedings of the Fourth International Workshop on Scalable Semantic Web knowledge Base Systems*.

48. Zaharia, Matei, Reynold S. Xin, Patrick Wendell, Tathagata Das, Michael Armbrust, Ankur Dave, Xiangrui Meng, Josh Rosen, Shivaram Venkataraman, Michael J. Franklin, Ali Ghodsi, Joseph Gonzalez, Scott Shenker, and Ion Stoica. 2016. Apache spark: A unified engine for big data processing. *Communications of the ACM* 59 (11): 56–65.

49. Zhang, Ying, Minh-Duc Pham, Óscar Corcho, and Jean-Paul Calbimonte. 2012. Srbench: A streaming RDF/SPARQL benchmark. In *The Semantic Web - ISWC 2012 - 11th International Semantic Web Conference, Boston, MA, USA, November 11–15, 2012, Proceedings, Part I*, 641–657.

Chapter 2
Preliminaries

Abstract This chapter offers a succinct overview of the background knowledge required to understand the content of the book. In particular, it introduces the basic notions behind streaming data processing as well as pillars for data integration using semantic technologies. In more specific terms, this chapter details the technologies apt to perform *selection* and *transformation* over streams of data, i.e., Data Stream Management Systems (DSMSs), which are opposed to traditional DataBase Management System (DBMS). Finally, this chapter deals with *integration* and *abstraction* operations, where the structure and technologies of the Semantic Web get described by including classic concepts like the Web of Data (WoD), Resource Description Framework (RDF), RDF Schema, the Web Ontology Language (OWL 2), the SPARQL query language and related semantics, and the Linked Data concept, with an updated possible depiction of the Semantic Web stack.

2.1 Introduction

This chapter positions the book within the state-of-the-art. It gives an overview of the concepts that are necessary to fully benefit from the remaining contents of the book. Furthermore, this chapter provides the reader with a strong basis to understand all theoretical aspects of Web stream analytics.

Within the Big Data context, a major conceptual foundation is provided by the definition of the "3 Vs," i.e., Volume, Variety, and Velocity [17]. The dimension of *Volume* relates to the scale of data sources, *Velocity* to the frequency of incoming data, and *Variety* to the heterogeneity of data coming from multiple sources [19]. These dimensions have a fundamental impact in determining which storage technologies, data pipelines, processing techniques, data quality features (e.g., structured data v. unstructured data), processing algorithms, and applications need to be selected and implemented, to achieve the expected performance from a Web stream data system.

These Big Data dimensions directly relate to the challenges of Web stream analytics[17], where data can reach the whole Web space (*Volume*), are produced at very high rates (*Velocity*), and come from heterogeneous sources (*Variety*).

The remainder of this chapter will go into more detail on the dimension of *Velocity* and *Variety*. We refer the reader to the survey of Rao et al. for an in-depth introduction to the dimension of *Volume* [21].

2.2 Data Velocity

This section introduces the background on data Velocity. It takes a deep dive on what *Data Streams* are and how they can be processed using *Continuous Queries*. The nuts and bolts of *Relational Stream Processing* will be discussed, explaining how the foundations of relational processing can be reused when processing relational streaming data. Lastly, this section explains how data systems behave w.r.t. time.

2.2.1 Data Streams and Continuous Queries

Stream processing is a subfield of data management that deals with *data-in-motion*, as opposed to the concept of *data-at-rest* which deals with static data persisting in a database or a certain file system.

The fundamental notion of *data-in-motion* is the concept of *data stream*, which identifies the continuous flow of an unbounded series of data through time. Figure 2.1 shows an example where the elements in the stream, also known as events, are colored boxes. Definition 2.1 gives a more formal definition of a *data stream*:

Definition 2.1 A data stream S is a possibly infinite multiset of elements $\langle s, \tau \rangle$, where s is a data item and $\tau \in \mathcal{T}$ is a timestamp of the element, with $\mathcal{T}$ a time domain (e.g., the set of natural numbers $\mathbb{N}$).

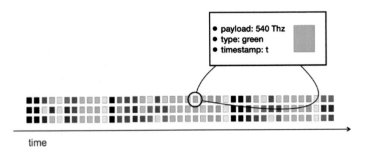

Fig. 2.1 Representation of a stream of codified colors

A Stream of Colors

Throughout the book, the color stream will be used to explain various concepts of Web stream analytics. As shown in Fig. 2.1, the color stream consists of elements $\langle s, \tau \rangle$, with s a colored box and τ a timestamp. In the figure, a green colored box has been highlighted. Its data item s describes the light spectrum as *payload* and the visible color as *type*. The timestamp for the highlighted box is t.

?   Why Is a Stream a Multiset of Elements?

A stream is defined as a multiset as in theory multiple elements with the same content could have the same timestamp. For example, in Fig. 2.1, there are multiple green colored boxes at the same time instance.

Stream Processing (SP) refers to a family of transformations executed over *data-in-motion*. In practice, SP requires to operate over the elements of one or more data streams. The unbounded nature of the data stream poses a significant challenge to the traditional methods of computation, as one cannot wait until the stream is finished and processes it as a whole. To address this issue, a wide range of methodologies, techniques, and tools capable of evaluating queries over an unbounded input has been proposed. In particular, two major kinds of transformations over data streams can be distinguished, i.e., *stateless* and *stateful*.

- *Stateless* transformations process one element of the stream at a time. Figure 2.2 shows an example, i.e., filtering out non-blue boxes from a stream of colors.
- *Stateful* transformations require to keep some state in memory, i.e., part of the data stream. Figure 2.3 depicts a *stateful* transformation that counts the number of boxes per minute. It includes the definition of a *window* to divide the stream into finite chunks from the unbounded stream.

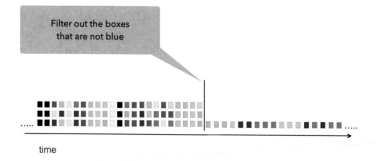

Fig. 2.2 Stateless stream processing: blue boxes are filtered out from the color stream

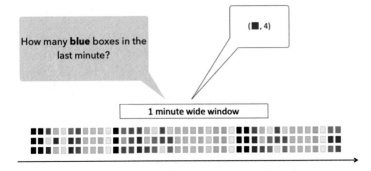

Fig. 2.3 Stateful stream processing: count each color in the last minute

Windows enable *stateful* transformations over unbounded data streams. They have the ability to extract a certain part of the *data-in-motion* and treat it as *data-at-rest* during a swift duration. This allows to perform transformations over the whole *extracted* dataset, i.e., the content of the window. Windows come in two flavors:

- Time-based window: extracts part of the stream based on temporal parameters, e.g., data produced in the last minute.
- Tuple-based window: extracts part of the stream based on the number of received data items, e.g., the last 100 received data items.

? Other Stateful Transformations?

Can you think of other transformations that require state, except for the aggregations explained above? Would joining two streams together require a window?

Querying data streams requires a special type of query that can handle data that are *in-motion* and thus continuously change. *Continuous queries* are a special class of queries that are registered only once and are then continuously executed over the data stream. Note that this is orthogonal to the idea of queries in traditional databases, where the data are fixed (or not changing that frequently) and the queries often change.

The idea of *continuous queries* is related to the concept of *continuous semantics*, i.e., processing an infinite input produces an infinite output [25]. It implies that the result of a *continuous query* is the set of results that would be returned if the query would be executed at every time instant.

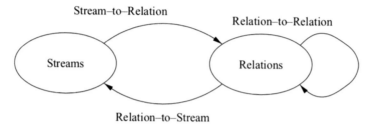

Fig. 2.4 CQL data abstractions from [4]

2.2.2 Relational Stream Processing

Arasu et al. introduce the Continuous Query Language (CQL) in [4]. CQL remains a pioneering framework for relational stream processing that inspired a number of languages and frameworks [7, 23]. They propose an abstract semantics for continuous queries, which build on two fundamental abstraction, i.e., *streams* and *relations*, which are depicted in Fig. 2.4. In practice, CQL allows to continuously process streams by extending the evaluation semantics that are well understood, e.g., Relational Algebra, so that it is possible to reuse the latter and still be able to efficiently process streams. To define continuous processing, a definition of time is mandatory:

Definition 2.2 The time τ is an infinite, discrete, ordered sequence of time instants $(\tau_1, \tau_2, ..., \tau_n)$, where $\tau_i \in \mathbb{N}$. A time unit is the difference between two consecutive time instants $(\tau_{i+1} - \tau_i)$, and it is constant.

Within this context of abstract semantics, a Time-Varying Relation (TVR), i.e., a relation that evolves over time, is a representation that alternatively can be regarded as a function mapping each point in time to a static relation. The foundational intuition with TVR is that both streams and tables can now be considered two representations of the same core semantic object. Before CQL, the accepted theory was that these two entities were separate, i.e., the stream/table duality, where a time-based event in SQL needed to be somehow materialized by different approaches. TVRs have been proposed as a common foundation for streaming SQL Declarative Language (DL). They are compatible with the classic principle of mutable database tables, and they bundle concepts like point-in-time queries, continuously updated views, and novel streaming queries. The whole set of basic SQL operators remains valid for time-varying relations, so that it results in maximal functionality reuse and low user's cognitive overhead. Formally, streams and relations are defined as follows [4]:

Definition 2.3 Let $\mathcal{T}$ be the ordered time domain, e.g., the set of natural numbers $\mathbb{N}$. A relational data stream S is a possibly infinite multiset of elements $\langle o, \tau \rangle$, where o is a data item, e.g., a tuple belonging to the schema of S, and $\tau \in \mathcal{T}$ is the timestamp of the element, e.g., a natural number.

Fig. 2.5 Representation of a
time-varying relation (TVR)

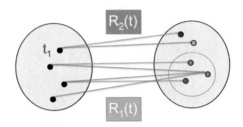

Definition 2.4 A relation R is a mapping from each time instant in τ to a finite but unbounded bag of tuples belonging to a fixed schema consisting of a set of named attributes.

This last definition specifies *time-varying relations*, where an instantaneous relation is the bag of tuples in a relation at a given point in time; therefore, given a relation R, $R(\tau)$ denotes an *instantaneous* relation (Fig. 2.5).

The core aspect of the CQL concept is that it describes stream-to-stream transformations as compositions of three families of operators that allow moving from stream to relation and vice versa, as shown in Fig. 2.4. They are specifically defined as:

1. *Stream-to-Relation (S2R)* operators that produce a relation from a stream;
2. *Relation-to-Relation (R2R)* operators that produce a relation from one or more other relations;
3. *Relation-to-Stream (R2S)* operators that produce a stream from a relation.

Within the definition context, while the R2R operators correspond to relational operators adapted to handle time-varying relations, the S2R operators in CQL are operators that cast a stream into *windows*, i.e., a time framed operator with a defined start point and a defined end point. As a consequence, a window is a set of elements extracted from a stream, denoted as W(S). CQL describes three specific typologies of it, as (i) *time-based*, (ii) *tuple-based*, and (iii) *partitioned* window operators. Within the context of the current analysis, the focus relies on the one expressed as *time-based* sliding window operator.

Definition 2.5 A time-based sliding window on a stream S takes a time interval I as a parameter and outputs a relation R of S [Range I] defined as:

$$R(\tau) = \{s | \langle s, \tau' \rangle \wedge \tau' \leq \tau \wedge \tau' \geq max\{\tau - I, 0\}\}$$

There are two special cases, formally defined by CQL in the following way:

1. $I = 0$ identified by RANGE [Now], $R(\tau)$ consists of tuples obtained from elements of S with timestamp τ
2. $I = \infty$ identified by RANGE [Unbounded], $R(\tau)$ consists of all the elements of S up to τ.

In CQL, the R2R operators correspond to relational operators adapted to handle time-varying relations, while the R2S operator family includes the *insert, delete,* and *relation* stream operators.

This brings forward the need for both a generic and a formal definition—according to the rules of Relational Algebra (RA)—for each and every one of them. The precondition is that of having two given consecutive time instants $\tau - 1$ and τ, so that the R2S operators can be defined as follows:

(1) The *insert stream* streams out all the new entries w.r.t. the previous instant, i.e.,

Definition 2.6 The insert stream applied to a relation R contains an element $\langle s, \tau \rangle$ if and only if the tuple s is in $R(\tau) - R(\tau - 1)$ at time τ:

$$Istream(R) = \bigcup_{\tau \geq 0} ((R(\tau) - R(\tau - 1)) \times \{\tau\}$$

(2) The *delete stream* streams out all the deleted entries w.r.t the previous instant, i.e.,

Definition 2.7 The delete stream applied to a relation R contains an element $\langle s, \tau \rangle$ if and only if the tuple s is in $R(\tau - 1) - R(\tau)$ at time τ:

$$Istream(R) = \bigcup_{\tau \geq 0} ((R(\tau - 1) - R(\tau)) \times \{\tau\}$$

(3) The *relation stream* streams out all the elements at a certain instant in the source relation, i.e.,

Definition 2.8 The relation stream applied to a relation R contains an element $\langle s, \tau \rangle$ if and only if the tuple s is in $R(\tau)$ at time τ:

$$Istream(R) = \bigcup_{\tau \geq 0} (R(\tau)) \times \{\tau\}$$

These three operator families allow defining the *continuous semantics* as follows:

Definition 2.9 (Continuous Semantics) Consider a query Q that is any type-consistent composition of operators from the above three classes. Suppose that the set of all inputs to the innermost (leaf) operators of Q is that of streams $S_1, \ldots, S_n$ ($n \geq 0$) and relations $R_1, \ldots, R_m$ ($m \geq 0$). The result of a continuous query Q at a time τ, which denotes the result of Q once that all inputs up to τ are available (a notion discussed below), is defined according to two cases:

- The outermost (topmost) operator in Q is of relation-to-stream type, producing a stream S. The result of Q at time τ is S up to τ, produced by recursively applying the operators comprising Q to streams $S_1, \ldots, S_n$ up to τ and relations $R_1, \ldots, R_m$ up to τ.

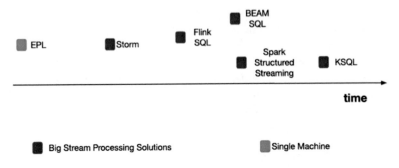

Fig. 2.6   A timeline of BigSPEs' SQL-like interface adoption [26]

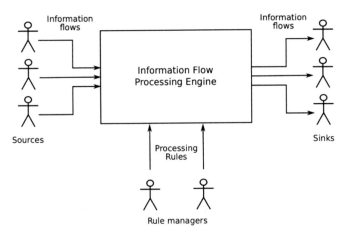

Fig. 2.7   Information flow processing engine [12]

- The outermost (topmost) operator in Q is of stream-to-relation or relation-to-relation type, producing a relation R. The result of Q at time τ is $R(\tau)$, produced by recursively applying the operators comprising Q to streams $S_1, \ldots, S_n$ up to τ and relations $R_1, \ldots, R_m$ up to τ (Fig. 2.6).

2.2.3   Stream Processing Engines

Cugola and Margara [12] proposed a first unifying view of stream processing systems, which they referred to as Information Flow Processor (IFP). Figure 2.7 shows a schematic representation. IFP users provide information needs that are registered and evaluated over input streams produced by multiple and distributed sources. Moreover, such information needs must be satisfied in a timely way, before the input data relevance expires. The authors' general model for an IFP engine builds on three main abstractions:

1. a set of *sources* generating information flows at the periphery of the system
2. a set of *rules* to filter, combine, and aggregate the different information flows
3. a set of *sinks* receiving the information flows.

Among other competing proposals and paradigms, two models in particular emerged, i.e., Data Stream Management System (DSMS) and Complex Event Processing (CEP). DSMS extends the traditional notion of Data Base Management System (DBMS). Indeed, the specialties of a DBMS are one-time queries and transformations over data-at-rest, while a DSMS rather reverses that interaction model, performing persisted queries over data-in-motion. This second approach relies on the notion of *continuous semantics* (see Definition 2.9). DSMSs were conceived starting from DBMSs, and, thus, they inherit both their data and query models [4]. CEPs are rule-based engines that aim at detecting events. An *event* is simply an atomic recording of "something that happens" within an unbounded stream of data. CEP is discussed in more detail in Sect. 2.2.4.

The contemporary BigSPE systems cover a wide variety of options when dealing with their engines' processing capabilities, e.g., monolithic v. distributed architecture or expressiveness v. simplicity of use w.r.t. declarative stream processing languages (Fig. 2.8).

There were some main assumptions taken with the design of Big Stream processing languages, and they can be summed up as general directions when dealing with their structure and usage. These languages are mainly developed to be (i) embedded within the Java programming language, which acts as a host-language; they are (ii) conceived so that developers are encouraged to explicitly

Fig. 2.8 BigSPE
programming stack [26]

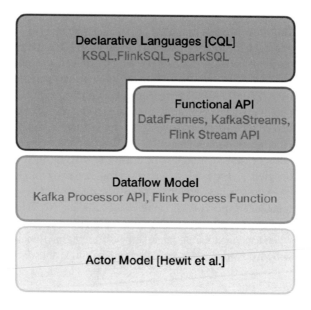

code according to the Directed Acyclic Graph (DAG) principle. Finally, they (iii) provide relational operators and—at the same time—they expose low-level details, e.g., partitioning or timestamp extraction.

Below, the main characteristics of distributed BigSPE systems and their relative DLs are reported. They bring some major advantages in terms of being horizontally, highly scalable, fault tolerant, implementing either *at-least-once* or *exactly-once* semantics, and they expose flexible APIs that guide toward the creation of DAGs. These systems share some major features, like (i) Windowing and Aggregates, (ii) Stream Enrichment (i.e., Stream-Table Joins), and (iii) Stream-to-Stream Joins. Next, the most prominent models used in Stream Processing are described:

Continuous Query Language (CQL) as reported at page 21 of Sect. 2.2.2. It is useful to remember that time in CQL refers to a logical clock tracking the evolution of relations and streams. Time is not expressed in the data being analyzed. Thus, in CQL, time is not represented as a first-class entity

SECRET Model proposed in the year 2010 by Botan et al. [8], whose cardinal concept is the *window*, i.e., a time-defined and time-spanned element, which explains the operational semantics of window-based SPEs. The model leverages upon four main stream-to-relation operators, as synthesized by Tommasini et al. [27], i.e., (i) the *Scope* function that maps an event time instant to the time interval over which the computation will occur, (ii) the *Content* function that maps a processing time instant to the subset of stream elements that occur during the event-time interval identified by the scope function, (iii) the *Report* dimension that identifies the set of conditions under which the window content becomes visible to downstream operators, and (iv) the *Tick* dimension that shows the conditions that cause a report to be evaluated.

Dataflow Model proposed in 2015 by Akidau et al. [2] presents a revolution in the approach to stream processing. Moreover, it lets the end user decide what degree of trade-off between correctness, latency, and cost should be considered. The processing model is based on streams of (*key, value*) pairs using two different primitives, i.e., (i) *ParDo*, for generic element-wise parallel processing producing zero or more output elements per input, and (ii) *GroupByKey*, for collecting data for a given key before sending them downstream for reduction. Such model addresses the problem of window completeness by stating that the mere technique of *watermarking* is not sufficient to regulate out-of-order arrivals in the stream. To compensate this deficiency, the Dataflow Model introduced two other key components, which appear to be complementary operators: (i) *Windows* that determine when data are grouped together for processing using event time and (ii) *Triggers* that determine when the results of groupings are emitted in processing time. The time window element, when used in streams, can assume different functional characteristics, i.e., *landmark* window, *tumbling* window, *hopping* window, and *sliding* window.

Stream and Table Duality, introduced by Sax et al. [23], leverages on three distinct notions: (i) *table* is the static view of an operator's state, updated for each input record and having a primary key attribute; (ii) *table changelog stream* is the dynamic view of the result of an update operation on a table, where the semantics of an update is defined over both keys and timestamps (replaying a table changelog stream allows to materialize the operator result as a table); and (iii) *record stream* represents facts instead of updates and it models immutable items, i.e., each record having a unique key; the last two are special cases of stream. All stream processing operators could be either *stateless* or *stateful* and might present single or multiple input streams.

2.2.4   Complex Event Processing

Complex Event Processing allows to detect temporal patterns over input data stream [15]. In the CEP state-of-the-art, several definitions for *events* have been proposed. In this book, the definition by Chakravarthy, which has been rephrased by Luckham [11, 18], will be used:

Definition 2.10   An *event e* is an atomic (happened completely or not at all) domain entity including a collection of facts, describing something that happened at a certain time [11].

Events can be considered as timestamped notifications of facts which are external to the detection system. The input data must be processed and interpreted according to defined detection rules which assign them a precise semantics. The computational models for CEP engines is typically based on Nondeterministic Finite-State Automata.

CEP engines evolved from *publish-subscribe* system [14]. Pub-Sub organizes the input information into topics and makes them one item at a time accessible to interested consumers via subscriptions and filters. CEP engines extend this behavior by including more expressive temporal operators to write the subscription rules.

Figure 2.9 shows the CEP model proposed in [12]. A CEP engine is often decentralized into a network of *agents* each of which is responsible for processing a subset

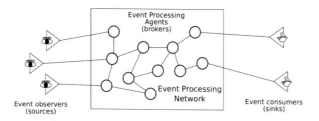

Fig. 2.9 CEP model proposed in [12]

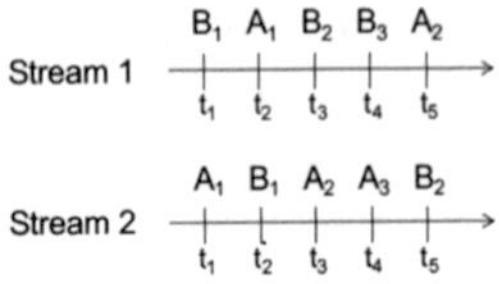

Fig. 2.10 Two example
streams to illustrate the
various event operators

of the input data detecting higher-level events and notifying subscriber. Typically,
events are specified at high level. Moreover, CEP languages allow composition
using time-aware operators. A list of prominent CEP operators includes, but is not
limited to the following ones. Notably, examples for each operators are provided
w.r.t. Fig. 2.10.

- *AND* is a binary operator: A AND B matches if both A and B occur in the stream
  and turns true when the latest of the two occurs in the stream. In Fig. 2.10, A
  AND B matches at t_2 in both stream 1 and stream 2.
- *OR* is a binary operator: A OR B matches if either A or B occurs in the stream.
  In Fig. 2.10, A OR B matches at t_1 in both stream 1 and stream 2.
- *SEQ* is a binary operator that takes temporal dependencies into account. A SEQ
  B matches when B occurs after A, in the time domain. In Fig. 2.10, A SEQ B
  matches at t_3 in stream 1 and at t_2 in stream 2.
- *NOT* is a unary operator: NOT A matches when A is not presented in the stream.
  NOT A matches at t_1 in stream 1 and t_2 at stream 2.
- *WITHIN* is a guard that limits the scope of the pattern within the time domain.
  A SEQ A WITHIN 2s matches in Fig. 2.10 at t_3 in stream 2 and not in stream
  1.
- *FIRST & LAST* are two unary operators. FIRST A (LAST A) matches when the
  first (last) A arrives in the stream. These operators typically require the use of
  the WITHIN guard to limit the evaluation scope.
- *EVERY* is a modifier that forces the reevaluation of a pattern once it has
  matched. EVERY A SEQ B matches at t_3 in stream 1 and at t_5 in stream 2
  for (A_2, B_2) and (A_3, B_2).

Note that more advanced temporal relations exist, such as the ones presented in
Allen's interval algebra [10].

2.3 Data Variety

This section presents the state-of-the-art for addressing Data Variety using Semantic
Web technologies. In particular, it discusses how to *combine* different data sources
(integration), as well as how to *represent* concepts through higher-level semantics
(abstraction).

The Semantic Web is an extension of the World Wide Web that aims at enabling
data interoperability by making data accessible to autonomous software agents. The
enabling of such machine-readable Web of Data (WoD) is achieved by the use of

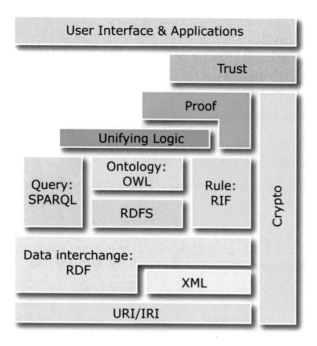

Fig. 2.11 Semantic Web stack of technologies [9]

ad hoc technologies to formally represent metadata, i.e., encoding semantics within the data. The current Semantic Web stack of technologies is depicted in Fig. 2.11. The organization that oversees the design, testing, and distribution of *standards*, viz., recommendations about Semantic Web technologies, is the World Wide Web Consortium (W3C), an entity founded in 1994 and led by Tim Berners-Lee, the inventor of the World Wide Web (WWW).

2.3.1   RDF and RDF Schema

The Resource Description Framework (RDF) is a semi-structured data model that fosters data interchange and allows publishing machine-readable representation of resources [28]. The basic structure of an RDF statement is the *triple*, consisting of *subject*, *predicate*, and *object*. Definition 2.11 formalizes this notion.

Definition 2.11 An RDF statement *t* is a triple

$$(s, p, o) \in (I \cup B) \times I \times (I \cup B \cup L)$$

where

- I is the set of IRIs
- B is the set of all the blank nodes, i.e., identifiers of anonymous resources
- L is the set of literals, i.e., string values associated with their data type
- I, B, and L are disjoint from each other.

RDF presents additional constructs that are relevant for integrating datasets. In particular, it is worth mentioning that a *set of RDF triples* is referenced to as an *RDF graph* (cf. Definition 2.12). Moreover, a *set of RDF graphs* is organized within datasets (cf. Definition 2.12).

Definition 2.12 An RDF graph is a set of RDF triples. An RDF dataset DS is a set:

$$DS = \{g_0, (u_1, g_1), (u_2, g_2), \ldots, (u_n, g_n)\}$$

where g_0 and g_i are RDF graphs, and each corresponding u_i is a distinct IRI. g_0 is called the default graph, while the others are called named graphs.

According to the W3C standard, the default serialization format for RDF is RDF/XML. Nevertheless, there are a plethora of other options when it deals with serialization (i.e., storing and transmitting data) for RDF-based semantic information, e.g., Turtle (TTL), JSON for Linked Data (JSON-LD), N-Triples Serialization Format (N-Triples), and TriG Serialization Format (TriG). Among those different formats, the formats that are known to be the most human-readable are Turtle and JSON-LD. Dealing with RDF triple statements, it is worth mentioning that the *subject* can either be a node or a blank node, the *predicate* is the relation that links a subject to an object, and the *object* can either be a node, a blank node, or a literal, e.g., a string value.

Turtle Syntax allows defining a Base URI with its relative Prefix for rendering the document in a more succinct way. In Listing 2.1, line 4 contains a triple with absolute IRIs. Line 1 declares a *base* IRI which is used by the triple in the following line. Line 2 declares a *prefix* IRI, which is used by the triple in the following lines. Also, Turtle syntax includes various shortcuts to represent RDF collections and literals. Turtle syntax allows grouping triples having the same subject by separating different predicates using a semicolon (cf. lines 10 and 11), while it allows separating alternative objects for the same predicate using a comma (cf. line 13 and 14). Blank nodes can be labeled (cf. Listing 2.1 line 16) or unlabeled (cf. Listing 2.1 line 18).

```
BASE <http://example/>
PREFIX p: <http://example/>

<http://example/subj1> <http://example/pred1> <http://example/obj1> .

<subj2> <pred2> <obj2> . # relative IRIs, e.g., http://example/subj2

p:subj3 p:pred3 p:obj3 . # prefixed name, e.g., http://example/subj3

p:subj4 p:pred4 p:obj4 ;
        p:pred5 p:obj5 .

p:subj5 p:pred6 p:obj6 ,
                p:obj7 .

_:bn1 p:pred7 _:bn2 .

[] p:pred8 [ p:pred9 "Literal" ] .
```

Listing 2.1 Example illustrating Turtle Syntax for Base and Prefix IRIs

According to the standard from Seaborn and Carothers [24], TriG is an extension of Turtle syntax that allows representing RDF datasets in a compact textual form. As shown in Listing 2.2, Trig provides syntactic shortcuts for graphs (cf. lines 6) and named graphs (cf. lines 11).

```
BASE <http://example.org/> .
PREFIX dct: <http://purl.org/dc/terms/> .
PREFIX foaf: <http://xmlns.com/foaf/0.1/> .

# default graph
{
   <bob> dct:publisher "Bob" .
   <alice> dct:publisher "Alice" .
}

<relations> {
   <alice> foaf:knows <bob> .
}
```

Listing 2.2 Example illustrating TriG syntax for RDF datasets

RDF Schema is a vocabulary for RDF data modeling and a semantic extension of RDF written itself in standard RDF, whose terms are described in the official specification [28]. It allows describing both groups of resources and the relationships between these resources. Classes and properties specified by RDF Schema are resources that are used to characterize other resources. RDF Schema describes properties in terms of the classes of the resource to which they apply. In this way, it is possible to define additional properties extending existing ones without the need to redefine the original description. RDF Schema refers to complementary specifications, among which it is worth naming the formal semantics of RDF that includes some entailment rules for reasoning, as indicated in Table 2.1.

Table 2.1 RDFS entailment patterns

RULE	If S contains:	then S RDFS entails recognizing D:
rdfs1	any IRI aaa in D	aaa rdf:type rdfs:data type .
rdfs2	aaa rdfs:domain xxx .	yyy rdf:type xxx .
	yyy aaa zzz .	
rdfs3	aaa rdfs:range xxx .	zzz rdf:type xxx .
	yyy aaa zzz.	
rdfs4a	xxx aaa yyy .	xxx rdf:type rdfs:Resource .
rdfs4b	xxx aaa yyy.	yyy rdf:type rdfs:Resource .
rdfs5	xxx rdfs:subPropertyOf yyy .	xxx rdfs:subPropertyOf zzz .
	yyy rdfs:subPropertyOf zzz .	
rdfs6	xxx rdf:type rdf:Property .	xxx rdfs:subPropertyOf xxx .
rdfs7	aaa rdfs:subPropertyOf bbb .	xxx bbb yyy .
	xxx aaa yyy .	
rdfs8	xxx rdf:type rdfs:Class .	xxx rdfs:subClassOf rdfs:Resource .
rdfs9	xxx rdfs:subClassOf yyy .	zzz rdf:type yyy .
	zzz rdf:type xxx .	
rdfs10	xxx rdf:type rdfs:Class .	xxx rdfs:subClassOf xxx .
rdfs11	xxx rdfs:subClassOf yyy .	xxx rdfs:subClassOf zzz .
	yyy rdfs:subClassOf zzz .	

2.3.2 Reasoning Techniques

In order to evaluate entailment regimes, two types of reasoning procedures can be used, i.e., Forward or Backward Chaining [22]. **Forward chaining** is the procedure of executing all the reasoning steps in order to infer all the possible derivations on the knowledge base [3]. This technique is often used to *materialize* the knowledge base, such that the reasoning procedure is executed once and multiple queries can be evaluated without any additional reasoning steps [1]. To explain the different reasoning paradigms, we will be using the color ontology as schema. It is a simple hierarchy stating the difference between *Cool* and *Warm* colors. Figure 2.12 visualizes such hierarchy.

Example 2.1 Forward chaining example. Let's assume the following rules are given in a knowledge base, extracted from the Color Ontology:

(R_1) :Blue *rdfs:subClassOf* :CoolColor
(R_2) :Green *rdfs:subClassOf* :CoolColor
(R_3) :Red *rdfs:subClassOf* :WarmColor
(R_4) :Brown *rdfs:subClassOf* :WarmColor
(R_5) :CoolColor *rdfs:subClassOf* :Color
(R_6) :WarmColor *rdfs:subClassOf* :Color

Fig. 2.12 A simple ontology
of colors

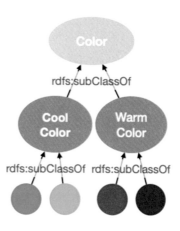

When reasoning about the triple <:blue$_1$, a, :Blue>, forward chaining would produce the following reasoning steps based on the rdfs9 rule that states that if A *rdfs:subClassOf* B and a$_1$ *rdf:type* A, then a$_1$ *rdf:type* B, it is possible to infer the following steps:

(I_1) :blue$_1$ *rdf:type* :CoolColor (based on rule R$_1$ and input triple)
(I_2) :blue$_1$ *rdf:type* :Color (based on rule R$_5$ and result of I$_1$)

The idea of *materialization* allows to execute a query, for example, to fetch all colors, without any need to execute the intermediate reasoning steps.

Backward chaining or query rewriting is the process of exploiting the knowledge base to find the data that are being queried [3]. No additional statements need to be inferred as with Forward Chaining; however, the reasoning is performed at query time.

Example 2.2 Backward Chaining Example.

Using the color ontology, when performing backward chaining or query rewriting for a query that looks for all Colors, the process will end up looking for all the possible subclasses of Color in the hierarchy. Backward chaining uses a top-down approach. When needing to find all the instances of the concept *:Color* (?c a :Color), and again having the input triple *(:blue$_1$, a, :Blue)*, the following backward steps would be performed:

(I_1) ?c a :CoolColor (based on rule R$_5$ and input query)
(I_2) ?c a :WarmColor (based on rule R$_6$ and result of I$_1$)
(I_3) ?c a :Red (based on rule R$_3$ and result of I$_2$)
(I_4) ?c a :Brown (based on rule R$_4$ and result of I$_2$)
(I_5) ?c a :Blue (based on rule R$_1$ and result of I$_1$)

I$_5$ would match the input triple *(:blue$_1$, a, :Blue)*. Note that when rewriting this process inside a query, it is necessary to inject all the possible options, thus resulting in many UNIONs. Each union can be evaluated independently and can be seen as an OR operator. Backward chaining would stop at I$_5$ as it matches the input triple; instead for query rewriting, it is necessary to evaluate all the possible options, resulting in an additional step:

(I_6)   ?c a :Green (based on rule R_2 and result of I_1)

When performing query rewriting, it is thus important that the whole hierarchy is represented in the rewritten query. Also, note that the given process specifically started from the queried concept *:Color*, even if there would be a parent concept of *:Color*.

2.3.3   Web Ontology Language

The Web Ontology Language (OWL) is a knowledge representation language for the *Semantic Web*. OWL became a W3C recommendation in 2004 and was updated in 2012. In this book, OWL always refers to OWL 2 unless otherwise specified. As a vocabulary, OWL builds on the RDF Schema Language, providing additional classes and properties to design ontologies for domains that present particular modeling challenges, e.g., the need for more expressiveness and complexity.

OWL enables *reasoning*, i.e., the possibility to infer implicit knowledge from asserted axioms. It relies on the theoretic semantics of Description Logic (DL), i.e., a family of well-studied knowledge representation languages [5]. OWL and OWL 2 are based on SHOIN(D) and SROIQ(D), respectively. Both languages are in the worst case logically intractable. Nonetheless, different fragments that present interesting computational properties are identified in the research literature. These are known as OWL 2 Profiles [29]. Figure 2.13 from [16] presents the relation

Fig. 2.13 OWL 2 profiles [16]

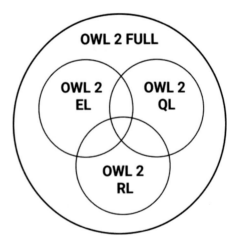

between OWL 2 Full and its profiles. Here follows a synthetic description of the main characteristics of the different OWL 2 profiles:

(C1) The **OWL 2 EL** profile allows performing basic reasoning in polynomial time w.r.t. the size of the ontology. It is designed for applications employing ontologies with a very large number of properties and/or classes and a small number of individuals. EL reflects the profile's basis in the EL family of Description Logics, providing only existential quantification [6].

(C2) The **OWL 2 QL** profile is designed for applications that use very large volumes of instance data and a small number of classes and properties. The name QL reflects the intent of designing a profile that privileges query answering. In particular, Conjunctive Query Answering (CQA) can be done using standard relational query languages, due to its low data complexity. OWL 2 QL is used in Information Integration systems that are based on Ontology-Based Data Access techniques [30].

(C3) The **OWL 2 RL** profile is designed for applications requiring scalable reasoning without sacrificing too much expressive power. Reasoning systems using this profile can be implemented using rule-based reasoning engines. Ontology consistency, class expression satisfiability, class expression subsumption, instance checking, and conjunctive query answering problems can be solved in polynomial time w.r.t. the ontology size.

2.3.4 SPARQL Protocol and RDF Query Language

SPARQL is the official Protocol and Query Language for interacting with RDF data, according to the official release of the standard by the W3C SPARQL 1.1 Community Group.[1] Portions of the specification that are relevant to the current analysis include SPARQL query language, SPARQL protocol, SPARQL service description vocabulary, and SPARQL entailment regimes. SPARQL vs. 1.1 extends the 1.0 version with additional features, like aggregations and sub-queries.

2.3.4.1 SPARQL Query Language

SPARQL Query Language, henceforth referred as SPARQL, is a graph query language for RDF data. Figure 2.14 presents the anatomy of a general query structure. A SPARQL query consists of three clauses, i.e., (a) Dataset Clause, (b) Query Form, and (c) Where Clause. The language also includes solution modifiers such as DISTINCT, ORDER BY, and GROUP BY. Finally, it is possible to associate a prefix label to an IRI. Here follows a thorough explanation of the SPARQL Query structure, as shown in Fig. 2.14.

---

[1] https://www.w3.org/TR/sparql11-query/.

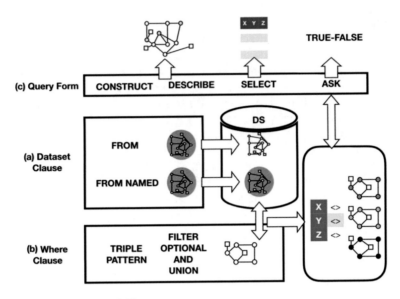

Fig. 2.14 Anatomy of a SPARQL query

SPARQL queries contain one (or more) *triple pattern*, which is the basic match unit in SPARQL. It is similar to an RDF triple, but it may contain *variables* in place of *resources*. A triple pattern gets the following formal definition.:

Definition 2.13 A triple pattern t_p is a triple (sp, pp, op) s.t.

$$(sp, pp, op) \in (I \cup B \cup V) \times (I \cup V) \times (I \cup B \cup L \cup V)$$

where I, B, and L are defined as in Definition 2.11, where V is the infinite set of variables.

```
PREFIX foaf: <http://xmlns.com/foaf/0.1/>              1
SELECT ?name ?email                                    2
FROM <http://www.w3.org/people>                        3
WHERE {                                                4
    ?person foaf:name ?name ;                          5
            foaf:age ?age .                            6
    FILTER (?age > 18)                                 7
}                                                      8
ORDER BY ?name                                         9
LIMIT 10                                               10
OFFSET 10                                              11
```

Listing 2.3 Example of a SPARQL query

The (a) Dataset Clause determines the RDF dataset in the scope of the query evaluation. This clause employs the FROM operator to load an RDF graph into the Dataset's default graph, while the FROM NAMED operator allows loading an RDF graph into a separated RDF graph identified by its IRI.

The (b) Where Clause allows defining the triple patterns to match over an RDF dataset. Listing 2.3 presents two triple patterns (cf. lines 5–6). A set of triple patterns is called a Basic Graph Pattern (BGP). BGPs can be named or not, and in the first case they are identified by either a *variable* or a *URI*. BGPs can include other compound patterns defined by different algebraic operators [20]. During the evaluation of a query, the graph from the dataset used for matching the graph pattern is called *Active Graph* and is set to the default graph of the dataset. Listing 2.3 shows an example of a FILTER clause that checks whether *?age* is greater than 18.

The (c) Query Form Clause allows specializing the query results. The SPARQL standard includes the following forms: (i) the **SELECT** form includes a set of variables and it returns the corresponding values according to the query results; (ii) the **CONSTRUCT** form includes a graph template and it returns one or more RDF graphs filling the template variables according to the query results; (iii) the **DESCRIBE** form includes a resource identifier and it returns an RDF graph containing RDF data that describe a resource; (iv) the **ASK** form returns a Boolean value signifying whether or not the query pattern has at least one solution.

2.3.4.2   SPARQL Evaluation Semantics

SPARQL semantics needs to be explicit, as well. A SPARQL query is defined as a tuple (E, DS, QF), where E is a SPARQL algebraic expression, DS an RDF dataset, and QF a query form.

Definition 2.14 A SPARQL query is defined as a tuple (E, DS, QF) where E is a SPARQL algebraic expression, DS an RDF dataset, and QF a query form.

The evaluation semantics of a SPARQL query algebraic expression w.r.t. an RDF dataset is defined for every operator of the algebra as $eval(DS(g), E)$ where E denotes an algebraic expression and $DS(g)$ a dataset DS with active graph g. Moreover, the evaluation semantics relies on the notion of solution mappings.

Definition 2.15 A solution mapping μ is a partial function

$$\mu : V \rightarrow I \cup B \cup L$$

from a set of variables V to a set of RDF terms.

A solution mapping μ is defined over a domain $dom(\mu) \subseteq V$, and $\mu(x)$ indicates the application of μ to a variable x.

Following the notation of Dell'Aglio et al. [13], Ω indicates a multiset of solution mappings and Ψ a sequence of solution mappings.

Definition 2.16 Two mappings, μ_1 and μ_2, are compatible, i.e., $\mu_1 \frown \mu_2$ iff:

$$\forall x \in (dom(\mu_1) \cap dom(\mu_2)) \Rightarrow \mu_1(x) = \mu_2(x)$$

Given an RDF graph, a SPARQL query solution is a set of solution mappings, each assigning terms of RDF triples in the graph to variables of the query.

The evaluation function $[\![.]\!]_D$ defines the semantics of graph pattern expressions. It takes graph pattern expressions over an RDF Dataset D as input and returns a multiset of solution mappings Ω. The evaluation over a Dataset D of triple pattern t and a compound graph pattern $P_i \ X \ P_j$ is defined as follows [20]:

$$[\![t]\!]_D = \{\mu | dom(\mu) = var(t) \in D\}$$

$$[\![P_1 \ AND \ P_2]\!] = \Omega_1 \bowtie \Omega_2 = \{\mu_1 \cup \mu_2 | \mu_1 \in \Omega_1 \wedge \mu_2 \in \Omega_2 \mu_1 \frown \mu_2\}$$

$$[\![P_1 \ UNION \ P_2]\!] = \Omega_1 \cup \Omega_2 = \{\mu_1 \cup \mu_2 | \mu_1 \in \Omega_1 \vee \mu_2 \in \Omega_2 \mu_1 \frown \mu_2\}$$

$$[\![P_1 \ OPTIONAL \ P_2]\!] = \Omega_1 \bowtie \Omega_2 = (\Omega_1 \bowtie \Omega_2) \cup (\Omega_1 / \Omega_2)$$

where

$$\Omega_1 / \Omega_2 = \{\mu | \mu \in \Omega_1 \wedge \nexists \mu' \in \Omega_2 \mu \frown \mu'\}$$

2.4   Chapter Summary

In this section, we explained how to handle both *Velocity* and *Variety* separately. In order to handle *Velocity*, we introduced the concepts of *Streams*, stateful/stateless *transformations*, and *windowing*. We learned about the CQL model, allowing to reuse well-known semantics of relation algebra on top of data streams. Furthermore, we peeked at the data models used within bit data stream processing engines. To handle *Variety*, we inspected the *Semantic Web* stack. We looked at *RDF* as a data model to enable data integration, RDFS, and OWL as means for defining vocabularies and SPARQL as a mean to query RDF data. We are now ready to investigate techniques that target both *Velocity* and *Variety* simultaneously.

References

1. Abiteboul, Serge, Richard Hull, and Victor Vianu. 1995. *Foundations of Databases*. Vol. 8. Boston: Addison-Wesley Reading.
2. Akidau, Tyler, Robert Bradshaw, Craig Chambers, Slava Chernyak, Rafael Fernández-Moctezuma, Reuven Lax, Sam McVeety, Daniel Mills, Frances Perry, Eric Schmidt, and Sam Whittle. 2015. The dataflow model: A practical approach to balancing correctness, latency, and cost in massive-scale, unbounded, out-of-order data processing. *PVLDB* 8 (12): 1792–1803.
3. Al-Ajlan, Ajlan. 2015. The comparison between forward and backward chaining. *International Journal of Machine Learning and Computing* 5 (2): 106.
4. Arasu, Arvind, Shivnath Babu, and Jennifer Widom. 2006. The CQL continuous query language: Semantic foundations and query execution. *The VLDB Journal* 15 (2), 121–142.
5. Baader, Franz, Ian Horrocks, and Ulrike Sattler. 2004. Description logics. In *Handbook on Ontologies. International Handbooks on Information Systems*, 3–28. Berlin: Springer.
6. Baader, Franz, Sebastian Brandt, and Carsten Lutz. 2005. Pushing the EL envelope. In *IJCAI*, 364–369. Professional Book Center.
7. Begoli, Edmon, Tyler Akidau, Fabian Hueske, Julian Hyde, Kathryn Knight, and Kenneth Knowles. 2019 One SQL to rule them all-an efficient and syntactically idiomatic approach to management of streams and tables. In *Proceedings of the 2019 International Conference on Management of Data*, SIGMOD '19, 1757–1772. New York, NY, USA: Association for Computing Machinery.
8. Botan, Irina, Roozbeh Derakhshan, Nihal Dindar, Laura Haas, Renée Miller, and Nesime Tatbul. 2010. Secret: A model for analysis of the execution semantics of stream processing systems. *PVLDB* 3: 232–243.
9. Bratt, Steve. 2007. Semantic web, and other technologies to watch. Online presentation.
10. Brown Jr., Allen L., Dale E. Gaucas, and Dan Benanav. 1987. An algebraic foundation for truth maintenance. In *IJCAI*, 973–980. Burlington: Morgan Kaufmann.
11. Chakravarthy, Sharma, and D. Mishra. Snoop: An expressive event specification language for active databases. *Data & Knowledge Engineering* 14 (1): 1–26.
12. Cugola, Gianpaolo, and Alessandro Margara (2012). Processing flows of information: From data stream to complex event processing. *ACM Computing Surveys (CSUR)* 44 (3): 15:1–15:62.
13. Dell'Aglio, Daniele, Emanuele Della Valle, Jean-Paul Calbimonte, and Óscar Corcho. 2014. RSP-QL semantics: A unifying query model to explain heterogeneity of RDF stream processing systems. *International Journal on Semantic Web and Information Systems (IJSWIS)* 10 (4), 17–44.
14. Eugster, Patrick Th., Pascal A. Felber, Rachid Guerraoui, and Anne-Marie Kermarrec. 2003. The many faces of publish/subscribe. *ACM Computing Surveys (CSUR)* 35 (2): 114–131.
15. Giatrakos, Nikos, Elias Alevizos, Alexander Artikis, Antonios Deligiannakis, and Minos Garofalakis. Complex event recognition in the big data era: A survey. *The VLDB Journal* 29 (1): 313–352.
16. Hoekstra, Rinke. 2009. *Ontology Representation - Design Patterns and Ontologies that Make Sense*. Vol. 197. *Frontiers in Artificial Intelligence and Applications*. Amsterdam: IOS Press.
17. Laney, Doug. 2001. 3d data management: Controlling data volume, velocity and variety. *META Group Research Note* 6 (70): 1.
18. Luckham, David. 2008. The power of events: An introduction to complex event processing in distributed enterprise systems. In *RuleML*. Vol. 5321. *Lecture Notes in Computer Science*, 3. Berlin: Springer.
19. Oussous, Ahmed, Fatima-Zahra Benjelloun, Ayoub Ait Lahcen, and Samir Belfkih. Big data technologies: A survey. *Journal of King Saud University-Computer and Information Sciences* 30 (4): 431–448.
20. Pérez, Jorge, Marcelo Arenas, and Claudio Gutiérrez. Semantics and complexity of SPARQL. *ACM Transactions on Database Systems (TODS)* 34 (3): 16:1–16:45.

21. Rao, T. Ramalingeswara, Pabitra Mitra, Ravindara Bhatt, and Adrijit Goswami. 2019. The big data system, components, tools, and technologies: A survey. *Knowledge and Information Systems* 60 (3): 1165–1245.
22. Russell, Stuart J., and Peter Norvig. 2009. *Artificial Intelligence: A Modern Approach.* Malaysia: Pearson Education Limited.
23. Sax, Matthias J., Guozhang Wang, Matthias Weidlich, and Johann-Christoph Freytag. 2018. Streams and tables: Two sides of the same coin. In *Proceedings of the International Workshop on Real-Time Business Intelligence and Analytics*, BIRTE '18. New York, NY, USA: Association for Computing Machinery.
24. Seaborne, Andy, and Gavin Carothers. 2014. RDF 1.1 trig. W3C recommendation, W3C. http://www.w3.org/TR/2014/REC-trig-20140225/
25. Terry, Douglas B., David Goldberg, David A. Nichols, and Brian M. Oki. 1992. Continuous queries over append-only databases. In *Proceedings of the 1992 ACM SIGMOD International Conference on Management of Data, San Diego, California, USA, June 2–5, 1992*, 321–330. New York: ACM Press.
26. Tommasini, Riccardo, Sherif Sakr, Marco Balduini, and Emanuele Della Valle. 2020. Tutorial: An outlook to declarative languages for big streaming data - DEBS 2020. http://streaminglangs. io
27. Tommasini, Riccardo, Sherif Sakr, Emanuele Della Valle, and Hojjat Jafarpour. 2020. Declarative languages for big streaming data. In *Proceedings of the 23rd International Conference on Extending Database Technology, EDBT 2020, Copenhagen, Denmark, March 30 –April 02, 2020*, ed. Angela Bonifati, Yongluan Zhou, Marcos Antonio Vaz Salles, Alexander Böhm, Dan Olteanu, George H. L. Fletcher, Arijit Khan, and Bin Yang, 643–646. OpenProceedings.org.
28. Wood, David, Markus Lanthaler, and Richard Cyganiak. 2014. RDF 1.1 concepts and abstract syntax. W3C recommendation, W3C. http://www.w3.org/TR/2014/REC-rdf11-concepts-20140225/.
29. Wu, Zhe, Ian Horrocks, Boris Motik, Bernardo Cuenca Grau, and Achille Fokoue. 2012. OWL 2 web ontology language profiles, 2nd ed. W3C recommendation, W3C. http://www.w3.org/ TR/2012/REC-owl2-profiles-20121211/.
30. Xiao, Guohui, Diego Calvanese, Roman Kontchakov, Domenico Lembo, Antonella Poggi, Riccardo Rosati, and Michael Zakharyaschev. 2018. Ontology-based data access: A survey. In *IJCAI*, 5511–5519. ijcai.org.

Chapter 3
Taming Variety and Velocity

Abstract Data on the Web are often very heterogeneous and volatile, a phenomenon that aligns with both Variety and Velocity in Big Data terms. Existing Big Data solutions have targeted Velocity through Stream Processing Engines and Variety through Semantic Web technologies. However, processing both Variety and Velocity simultaneously is still an open research problem. This chapter explains how Variety and Velocity can be targeted at once through RDF Stream Processing (RSP) in order to process streams on the Web. RSP focuses on the evaluation of Semantic Web technologies over RDF data streams. This chapter presents the fundamental notions of RDF Streams, Web Events, and the RSP-QL language family. Theoretical foundations of RSP will be detailed, allowing to understand the internals and challenges of RSP engines. Next to various examples of RSP engines, efficient inference schemes will be detailed that allow to perform reasoning over RDF data streams. This chapter provides ample examples and details how different types of analytics can be employed over Web Streams.

3.1 Introduction

This chapter discusses how Web streams can be processed. Web streams are both heterogeneous, as are most Web data, and volatile. This relates to Variety and Velocity in Big Data terms [24]. Existing Big Data solutions have targeted Velocity through Stream Processing Engines [15, 21], while the problem of Variety has been targeted through Semantic Web technologies [14]. However, processing both Variety and Velocity simultaneously is a challenging problem [11, 29]. RDF Stream Processing (RSP) is an upcoming field of research that aims at solving both data Variety- and Velocity-related processes, combined through the principle of continuous evaluation, offered by Semantic Web technologies over RDF streams [12].

This chapter focuses on query answering over RDF streams and details theoretical foundations of RSP while providing ample practical examples. Next to the foundations of RSP, this chapter introduces RSP4J, an API for the development of RSP engines. Furthermore, this chapter details how RSP engines or applications can be created through RSP4J. This chapter also describes the challenges of performing inference over RDF streams and details efficient reasoning algorithms. The *color stream* from previous chapters is used to explain the concepts of RSP. The following sections are built on the knowledge over RDF and SPARQL semantics introduced in Chap. 2. Let's first dive in into theoretical foundations of RSP.

3.2 RDF Stream Processing

In the realm of RSP, a variety of RSP languages emerged over time, e.g., C-SPARQL [5], CQELS-QL [16], SPARQL$_{stream}$ [7], and Strider-QL [22]. Such languages were extensions of SPARQL and supported some form of continuous semantics. RSP languages usually paired with working prototypes that helped proving the feasibility of the approach as well as studying its efficiency. However, each of these prototypes had its own internal execution semantics making them hard to compare. To resolve this issue, a reference model was proposed, i.e., RSP-QL [10], that unified existing RSP dialects and the execution semantics of existing RSP engines.

The next section explains the RSP-QL reference model, in order to understand the internal of RSP engines.

3.2.1 From CQL to RSP-QL

In the literature, there are many definitions of data streams, with a general agreement on considering them as unbounded sequences of time-ordered data. Different notions of time are relevant for different applications. The most important ones are the time at which a data item reaches the data system (*processing time*) and the time at which the data item was produced (*event time*) [2]. In RSP, streams are represented as RDF objects, as stated by Definition 3.1 [10].

Definition 3.1 A data stream S is an infinite sequence of tuples $\langle d_i, t_e, t_p \rangle$ where d_i is a data item and t_e/t_p are, respectively, the event time and the processing time timestamps. An **RDF Stream** is a stream where the data item d_i is an RDF object and t_e/t_p are timestamps indicating event time and processing time, respectively.

Operationally, stream processing requires a special class of queries, as they need to be constantly evaluated as the data change continuously. This class of queries runs under *continuous semantics*, as defined in Definition 3.2.

Definition 3.2 Under continuous semantics, the result of a query is the set of results that would be returned if the query was executed at every instant in time.

In practice, continuous queries consume one or more infinite inputs and produce an infinite output [26]. Arasu et al. [4] proposed CQL, a query model for processing relational streams based on three families of operators, as depicted in Fig. 3.1.

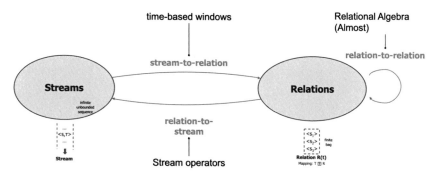

Fig. 3.1 The CQL query model, i.e., the S2R, R2R, and R2S operators

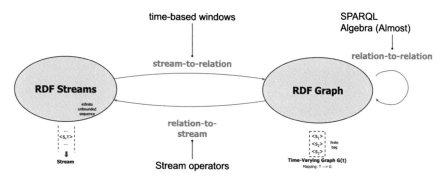

Fig. 3.2 The RSP-QL query model, i.e., the S2R, R2R, and R2S operators

The basic notions of CQL have been introduced in Chap. 2. As proposed by the RSP-QL semantics, the CQL model is extended to enable the processing of RDF streams. The adaptation of the CQL model according to the RSP-QL semantics is visualized in Fig. 3.2.

Stream-to-Relation (S2R) is a family of operators that bridges the world of streams with the world of relational data processing. These operators chunk the streams into finite portions. A typical operator of this kind is a Time Window operator. In RSP-QL, a time-based window operator is defined as in Definition 3.3.

Definition 3.3 The time-based window operator $\mathbb{W}$ is a triple (α, β, t^0) that defines a series of windows of width (α) and a slide of (β), starting at t^0.

Relation-to-Relation (R2R) is a family of operators that can be executed over the finite stream portions. In the context of RSP-QL, R2R operators are SPARQL 1.1 operators evaluated under continuous semantics.

To clarify this intuition, Dell'Aglio et al. introduce the notion of a Time-Varying Graph and RSP-QL dataset [10]. A **Time-Varying Graph** is the result of applying

Fig. 3.3 Visualization of the time-varying graph as a function of time

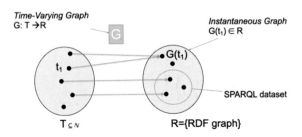

a window operator $\mathbb{W}$ to an RDF Stream S (see Definition 3.4), while the RSP-QL dataset (SDS) is an extension of the SPARQL dataset for continuous querying (see Definition 3.5).

Definition 3.4 A **Time-Varying Graph** is a function that takes a time instant as input and produces as output an RDF graph, which is called an *instantaneous graph*.

Given a window operator $\mathbb{W}$ and an RDF stream S, the Time-Varying Graph $TVG_{\mathbb{W},S}$ is defined where the $\mathbb{W}$ is defined.

Figure 3.3 visualizes the Time-Varying Graph function with a time instant ($\in T$) as its domain, producing an RDF graph as output.

In practice, for any given time instant t, $\mathbb{W}$ identifies a subportion of the RDF stream S containing various RDF graphs. The Time-Varying Graph function returns the union (*coalescing*) of all the RDF graphs in the current window.[1]

Definition 3.5 An **RSP-QL dataset** SDS extends the SPARQL dataset[2] as follows: an optional default graph A_0, n ($n \geq 0$) named Time-Varying Graph and m ($m \geq 0$) named sliding window over k ($k \leq m$) data streams.

An RSP-QL query is continuously evaluated against an SDS by an RSP engine. The evaluation of an RSP-QL query outputs an instantaneous multiset of solution mappings for each evaluation time instant. The RSP engine's operational semantics determines the set ET of evaluation time instants.

Finally, *Relation-to-Stream* (R2S) is a family of operators that returns a set of infinite data from a finite dataset, allowing to stream out the results. RSP-QL includes three R2S operators:

1. the **RStream** that emits the current solution mappings
2. the **IStream** that emits the difference between the current solution mappings and previous ones
3. the **DStream** that emits the difference between the previous solution mappings and the current ones.

Figure 3.4 shows an example of the different operators in action. At the bottom, the Stream S is depicted that streams the colors from the running example. The

---

[1] The current window identified by $\mathbb{W}$ with the oldest closing time instant at t.

[2] https://www.w3.org/TR/rdf-sparql-query/#specifyingDataset.

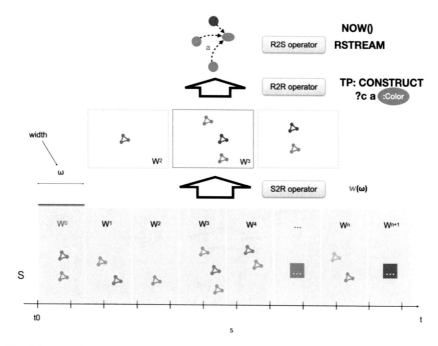

Fig. 3.4 Example for the S2R, R2R, and R2S operators

stream gets chunked into windows using the S2R operator. The R2R operator evaluates a SPARQL CONSTRUCT query consisting of a single Triple Pattern (TP) that extracts and filters out all the colors and creates a color graph. The R2S operator uses RStream to stream out all the color graphs.

Figure 3.5 goes a step further and shows the different behavior for all three R2S operators. In the depicted example, the S2R and R2R operators behave similarly as in Fig. 3.4, with the difference that now, instead of W^3, window W^4 is active. To illustrate the behavior of the different R2S operators, the color graph of the current window (W^4) needs to be compared to the ones from the previous window (W^3). In the current graph, there is one red and one green result, while in the previous window there were two blue and one red. For the *RStream*, the current result can be emitted, without comparing it against any of the previous results. The DStream emits that in the current result, none of the blue results are present anymore, while the IStream emits that there are now green results.

> **Can you define the output of the R2S operators when W^5 triggers?**

As an exercise, can you define the output of the RStream, IStream, and DStream when the next window (W^5) triggers? You can look at Fig. 3.5 to see the time-varying graph of the previous window (W^4) and create the resulting color graph

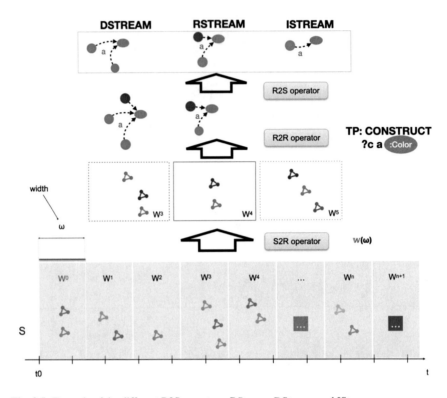

Fig. 3.5 Example of the different R2S operators: RStream, DStream, and IStream

for the current window (W^5), knowing there is a pink, green, and red color in the window.

❗ Format of the Output

It is important to note that the format of the output stream depends on the query form used in the R2R operator and is thus not always an RDF stream. When using the *Select* query form, the output is a stream of solution mappings; indeed, only when using a *Construct* (or *Describe*) query form the output is an RDF stream.

❓ What Happens When No R2S Operator Is Defined?

When no R2S operator is defined, the output is not a stream. The output is directly the result of the R2R operator and is thus a *relation* that changes over time.

3.2.2 RSP-QL Query Language

Even though there has been no formal proposal for the syntax of the RSP-QL query language, the RSP community has come to an agreement over a fixed syntax. The SPARQL syntax has been extended with a (named) *WINDOW* clause for the definition of the S2R and a *REGISTER* clause for the definition of the R2S. The *WINDOW* clause allows to define the name of the window, such that it can be used in the WHERE clause of the query, detailing on which stream the S2R needs to operate, plus the width and slide parameters of the window. The *REGISTER* clause can define any kind of R2S (IStream, DStream, or RStream) and name the output stream. Listing 3.1 shows the example of an RSP-QL query. Line 4 defines the *REGISTER* clause, while line 6 defines the window definition through the *WINDOW* clause. On line 8, the window is defined that needs to be targeted in the *WHERE* clause.

```
1   PREFIX color: <http://linkeddata.stream/ontologies/colors#>
2   PREFIX :      <http://linkeddata.stream/resource/>
3
4   REGISTER RSTREAM :outStream AS
5   SELECT ?green
6   FROM NAMED WINDOW :window ON :colorStream [RANGE PT15S STEP PT5S]
7   WHERE {
8     WINDOW :window {
9        ?green a color:Green.
10    }
11  }
```

Listing 3.1 RSP-QL query with register and window clauses defining the R2S and S2R operators

Figure 3.6 shows the extension of the SPARQL architecture from Chap. 2 with the *Window* clause that populates the SDS with time-varying graphs.

3.2.3 Reporting Strategies

After looking at the different operators (S2R, R2R, and R2S), the question remains when RSP engines should start consuming the content of the windows defined through the S2R operators. *Reporting Strategies* define the conditions under which the engine emits the content of the window. In general, there can be four different reporting strategies:

- **(CC)** *Content Change*: the engine reports when the content of the current window changes
- **(WC)** *Window Close*: the engine reports when the current window closes
- **(NC)** *Non-empty Content*: the engine reports when the current window is not empty

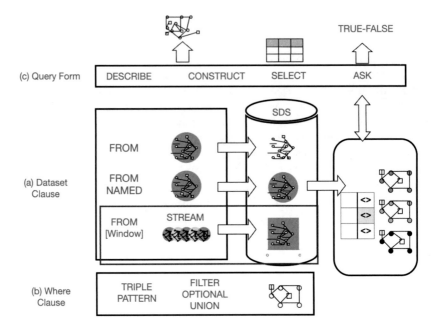

Fig. 3.6 RSP-QL query architecture

- **(P)** *Periodic*: the engine reports periodically.

Note that combinations of reporting dimensions are possible.

3.3   Reasoning over Web Streams

In order to infer implicit facts, align data from different schemas or simplify query definitions; deductive reasoning can be very useful. Reasoning techniques can be split up in two groups [3]: *forward chaining* techniques and *backward chaining* (or query rewriting) techniques [23]. Each of them is suited for specific use cases, but none of them are directly suited for reasoning over Web streams [6]. To explain the different reasoning paradigms, the color ontology is used as schema: it is a simple hierarchy stating the difference between *Cool* and *Warm* colors. Figure 3.7 visualizes such hierarchy. The hierarchy in Fig. 3.7 differentiates between *Primary* and *Secondary* colors, both for *Cool* and *Warm* colors.

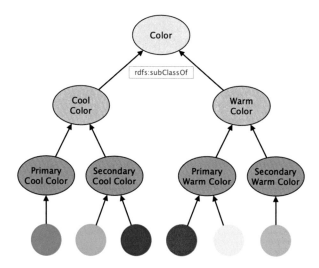

Fig. 3.7 Graphical representation of the color ontology

3.3.1   Problems with Standard Reasoning Techniques

3.3.1.1   Forward Chaining

As detailed in Sect. 2.3.2, forward chaining is the procedure of executing all the reasoning steps in order to infer all the possible derivations on the knowledge base. This technique is often used to *materialize* the knowledge base, such that the reasoning procedure is executed once and multiple queries can be evaluated without any additional reasoning steps.

When reasoning over the triple <:blue$_1$, a, :Blue>, forward chaining would produce the following inferred triples taken the color hierarchy from Fig. 3.7 into account:

(I_1)   <:blue$_1$, a, :PrimaryCoolColor>
(I_2)   <:blue$_1$, a, :CoolColor>
(I_3)   <:blue$_1$, a, :Color>

Materialization computes all derivation steps and explicitly stores them, allowing to execute queries, for example, to fetch all colors, without needing to perform any intermediate reasoning step. Listing 3.2 shows the RSP-QL query that fetches all colors in the color stream.

```
1 PREFIX color: <http://linkeddata.stream/ontologies/colors#>
2 PREFIX :       <http://linkeddata.stream/resource/>
3 SELECT ?c
4 FROM NAMED WINDOW <cw> ON :colorstream [RANGE PT15S STEP PT5S]
5 WHERE {
6 WINDOW ?cw {
7 ?c a color:Color .
8 }
9 }
```

Listing 3.2 An example query based on the Ontology hierarchy

Even though only *Colors* are of interest for the evaluation of the query, there have been triples inferred that cannot contribute to the answering of the query, i.e., I_1 and I_3. The derivation of inferred triples that do not contribute to the evaluation of the continuous queries is a typical problem for forward chaining reasoning techniques.

The forward chaining technique is fine in cases where data do not change often; however, all the available resources should be employed as efficiently as possible in streaming scenarios. Furthermore, through the evaluation of *Continuous Queries*, it is known which queries will be evaluated in advance. Even from the above simple example, it is possible to see that unnecessary triples are inferred, unnecessary since the query is only interested in fetching the different colors. This results in wasted CPU time and memory consumption, ultimately reducing throughput [6].

3.3.1.2 Backward Chaining

Backward chaining or query rewriting is the process of exploiting the knowledge base to find the data that are being queried. The advantage of this technique is that there is no additional memory or storage required to store the additional inferred statement as required with Forward Chaining. The drawback is that the reasoning is performed at query time, thus lowering performance when multiple queries need to be evaluated.

Using backward chaining or query rewriting results in queries that include many UNION statements. Listing 3.3 shows the RSP-QL query that has been rewritten in order to query all the *Colors* (?c a color:Color). This query is the result of injecting the knowledge from the ontology hierarchy into the query itself. The more UNION statements a query contains, the longer it will take to be evaluated [6]. Note that our color ontology is rather simple: using this technique with larger ontologies quickly results in queries with an extremely larger number of UNION statements, forcibly slowing evaluation times. The latter is an outcome that should be avoided at all costs in streaming scenarios.

```
1   PREFIX color: <http://linkeddata.stream/ontologies/colors#>
2   PREFIX :       <http://linkeddata.stream/resource/>
3   SELECT ?c
4   FROM NAMED WINDOW <cw> ON :colorstream [RANGE PT15S STEP PT5S]
5   WHERE {
6   WINDOW ?cw {
7   {?c a color:Color .}
8   UNION {?c a color:CoolColor}
9   UNION {?c a color:WarmColor}
10  UNION {?c a color:PrimaryCoolColor}
11  UNION {?c a color:PrimaryWarmColor}
12  UNION {?c a color:SecondaryCoolColor}
13  UNION {?c a color:SecondaryWarmColor}
14  UNION {?c a color:Blue}
15  UNION {?c a color:Green}
16  UNION {?c a color:Red}
17  UNION {?c a color:Yellow}
18  UNION {?c a color:Orange}
19  UNION {?c a color:Violet}
20
21  }
22  }
23
```

Listing 3.3 An example of query rewriting based on the Ontology hierarchy

3.3.2 Efficient Hierarchical Reasoning with C-Sprite

C-Sprite [6] is an efficient algorithm for reasoning over hierarchies of concepts, such as our Color hierarchy. It exploits the idea that queries in RSP are registered only once and they are continuously evaluated. Thus, the queries do not change very often. C-Sprite exploits this idea to optimize reasoning steps, for hierarchical reasoning, based on the registered queries. In this context, hierarchical reasoning is the task of computing *rdf:subClassOf* relations, which resemble a hierarchy of concepts.

Conceptually, C-Sprite prunes the class hierarchy and eliminates all reasoning steps that are of no interest for the registered queries. Remember that when performing forward chaining in Sect. 3.3.1.1, the focus was exclusively centered on all *Colors*; however, the reasoning process also inferred *CoolColors* and *PrimaryCool-Color*, which were not of interest to the query, and thus these inferred concepts were eventually dropped. When the hierarchies become very large, this becomes a problem in terms of computational performance. Therefore, C-Sprite only infers facts that might result in the evaluation of the registered queries. Figure 3.8 shows a conceptual representation of how the hierarchy is pruned when querying all *WarmColors*. Note that the *PrimaryWarmColor* and *SecondaryWarmColor* concepts are not dropped, as in theory a *WarmColor* might arrive in the stream as well.

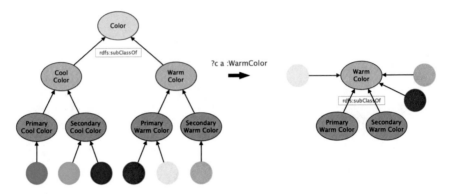

Fig. 3.8 Conceptual representation of C-Sprite hierarchy pruning, based on color query

Color and the *CoolColor* sub-hierarchy are dropped as they are not subclasses of *WarmColor*.

3.3.2.1   C-Sprite Under RSP-QL

Remember that RSP-QL consisted of S2R, R2R, and R2S operators. A typical RSP-QL program can be written as a pipeline of all three operators:

$$RSP - QL = S2R + R2R + R2S \tag{3.1}$$

C-Sprite can be seen as an extension of the R2R operator that performs hierarchical reasoning next to the evaluation of the query:

$$RSP - QL_{csprite} = S2R + R2R_{csprite} + R2S \tag{3.2}$$

As a matter of fact, with C-Sprite, it is possible to perform a special step before the evaluation of the query. First the reasoning steps get performed, and then the query is evaluated:

$$RSP - QL_{csprite} = S2R + CSprite(R2R) + R2S \tag{3.3}$$

As long as C-Sprite does not implement any joins, and it focuses on hierarchical reasoning, it can in theory be executed over a single event in the stream. This allows to perform the reasoning right away but also allows to filter out any triples that are not of interest to the evaluation of the query:

$$RSP - QL_{csprite} = S2R_0 + CSprite + R2S_0 + S2R + R2R + R2S \tag{3.4}$$

with $S2R_0$ being the window operator that takes one element from the stream and $R2S_0$ the operator that streams out the results of the reasoning.

3.3.2.2 A Data Structure for Efficient Hierarchical Reasoning

For efficient lookup of the parent classes for a specific class in the hierarchy, a possible data structure is obtainable by saturating the hierarchy and storing for each class a list of all the parents, as visualized in Fig. 3.9a. By storing the list of parents in a hashmap, using the class name as the key and storing the list of parents as the value, one can look up the parents for a specific class in constant time ($O(1)$).

Since a way to efficiently query the data is needed, a new instance of the hierarchy is created that only contains the concepts (keys) that have the queried type in their list of parent concepts. When the concepts have been filtered, each concept is linked to the query. When multiple queries are added, each concept contains a list of queries it matches according to the hierarchy. C-Sprite focuses specifically on queries asking for specific type instances (queried types). In Fig. 3.9, query Q1 asks for all the instances of *WarmColors*. The concepts that are not *WarmColors*, such as *CoolColors* (and subclasses) and *Colors*, are dropped. For the other concepts, a direct link is made to the query.

When new data arrive, such as a *blue* in Fig. 3.9, a simple lookup in the hashmap allows to detect that query Q1 matches.

3.3.2.3 C-Sprite Algorithm

Let's now define an algorithm to query hierarchical classes encoded on the data structures introduced above. Algorithm 1 shows the pseudo-code that describes how the data structure is constructed to efficiently perform the querying. First, the ontology hierarchy gets converted into a hashmap H containing all its parents for each class. Each time a query is registered, a copy of the hierarchy is pruned such that it only contains the concepts that have the queried type in their list of parents. The selected concepts are then directly linked to the queries. This allows to perform the hierarchical reasoning as a lookup in a hashmap.

Algorithm 2 is executed on the ingestion of a new *triple*. When a new triple is received, the system executes the *CheckHierarchyMatch* function that takes the triple and the pruned hierarchy hashmap as arguments. By looking up the asserted types of the triple in the hashmap, it detects which queries the triples match.

Complexity Study Let's assume that the number of queried classes in all the queries is m (i.e., $m = \sum_{i=0}^{len(Q)} len(Q_i)$). Thus, the complexity of first looking up in the pruned hashmap if the triple's type matches any queries and then iterating over them is $O(1)+O(m)$. The complexity only depends on the number of queries, which is typically low. Indeed, for each triple in the stream, the execution is performed in constant time.

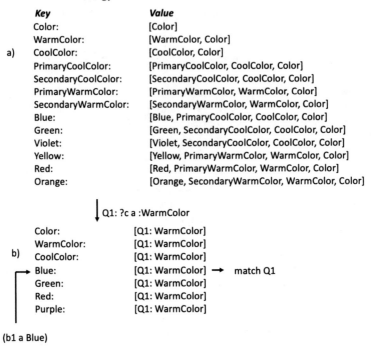

Fig. 3.9 Pruning of hierarchy and storing of concepts using C-Sprite

---

Algorithm 1 Query registering
***
Require: Q a collection of queries, each interested in one or more types.
  1: $H \leftarrow ConvertToHierarchy(O)$ ▷ Stores parents for each class in the Ontology O
  2: **function** PREPAREHIERARCHY(H, Q)
  3:     $H' \leftarrow []$
  4:     **for** $q \in Q$ **do**
  5:         **for** $(concept, parents) \in H$ **do**
  6:             **if** $q \in parents$ **then**
  7:                $H'[concept].append(q)$
  8:             **end if**
  9:         **end for**
10:     **end for**
11:     **return** H'
12: **end function**

3.3.3   Incremental Maintenance Approaches

When computing transitive properties, or inference of rules as found in OWL2 RL [17] using forward chaining, some kind of maintenance approach [1] is necessary to update the materialization when data is added or removed. Removal

---

Algorithm 2 Calculate the query matches on a hierarchical level

---

Require: Q a collection of queries, each interested in one or more types.
1: $H \leftarrow ConvertToHierarchy(O)$ ▷ Stores parents for each class in the Ontology O
    (preprocessing step)
2: $H' \leftarrow PrepareHierarchy(H, Q)$ ▷ (preprocessing step)
3: $triple \leftarrow$ ClassAssertion(type,subject)
4: **function** CHECKHIERARCHYMATCH(H', $triple$)
5:      $QueryMatches \leftarrow H'(types(triple))$ ▷ types extracts the type assertions of a triple
6:      **return** $QueryMatches$
7: **end function**

---

is especially tricky, as multiple statements can cause the same statements to be deduced, making it often unclear when the inferred statements do not longer hold. Various approaches exist to incrementally maintain this materialization [18]:

- Counting-based approaches [13] maintain a counter that is incremented every time the same statement is inferred. The counter is deducted when one of the facts that led to the deduction is removed. This however requires to sacrifice memory as multiple counters need to be maintained
- Delete/re-derive algorithm does not require any additional bookkeeping as the count-based approach [8, 25]. It overdeletes facts and then tries to figure out which deduction can be re-derived from other statements and inserts them again
- Forward/backward/forward algorithm [19] uses forward chaining to overdelete and then uses backward chaining to validate if each of the facts still holds, making it more efficient than the Delete/re-derive algorithm. Forward/backward/forward algorithm is also used in the popular RDF store RDFox [20].
- IMaRS [9] is an incremental maintenance approach tailored specifically for window-based approaches, as deletion can be foreseen when statements go out of the bounds of the window. An expiration time annotation can be added to all the statements involved in the materialization, allowing to identify which deductions need to be deleted when the window shifts.

3.4   RSP4J: An API for RSP Engines

This section presents RSP4J [28], an API for the development of RSP engines under RSP-QL semantics. Figure 3.10 shows RSP4J core modules, i.e., (a) Querying, (b) Streams, (c) Operators, (d) the SDS, and (e) the Engine with Execution Semantics. To provide concrete examples of RSP4J, Yet Another Stream Processing Engine for RDF (YASPER) is used. YASPER is a strawman proposal[3] designed for teaching purposes.

---

[3] https://en.wikipedia.org/wiki/Straw_man_proposal.

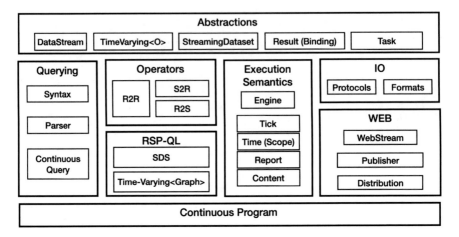

Fig. 3.10 Blocks of the RSP4J's modules by type

3.4.1 Querying

```
PREFIX : <http://example.org#>
REGISTER RSTREAM <output> AS
SELECT AVG(?v) as ?avgTemp
FROM NAMED WINDOW :w1 ON STREAM :stream1 [RANGE PT5S STEP PT2S]
WHERE {
    WINDOW :w1 { ?sensor :value ?v ; :measurement: ?m }
    FILTER (?m == 'temperature')
}
```

Listing 3.4 An example of RSPQL query

The query module contains the elements for writing RSP-QL programs in a declarative way. The syntax is based on the proposal by the RSP community. At this stage of development, RSP4J accepts SELECT and CONSTRUCT queries written in RSP-QL syntax (e.g., Listing 3.4).[4] Although RSP-QL [10] does not discuss how to handle multi-streams, RSP4J does, allowing its users to fully replicate the behavior of existing systems.

Moreover, RSP4J includes the ContinuousQuery interface that aims at making the syntax extensible. Indeed, RSP4J users can bypass the syntax module and programmatically define extensions in the query language.

---

[4] The RSP W3C community group has started working toward a common syntax and semantics for RSP (https://github.com/streamreasoning/RSP-QL).

3.4.2   Streams

The *Streams* module allows providing your own implementation of a data stream. It consists of two interfaces inspired by VoCaLS [27]: the WebStream and WebDataStream.

WebStream represents the stream as a Web resource, while WebDataStream represents the stream as a data source. Figure 3.11 provides an overview of the relationships across these classes and interfaces.

The WebStream does not include any particular logic. It is identified by an HTTP URI so it can be de-referenced and then consumed through an available endpoint [27].

Listing 3.5 shows the implementation of the WebDataStream interface, which exposes two methods: put and addConsumer.

The former allows injection of timestamped data items of type *E* by producers; the latter connects the stream to interested consumers, e.g., window operators, or super-streams. The interface is generic, and it allows RSP4J's users to utilize multiple RDF Stream representations, i.e., either RDF Graphs or Triples, or even non-RDF Web Streams. A WebDataStream might also include some metadata relevant for the processing, i.e., links to ontologies, SHACL schemas, or alternative endpoints.

```
public class RDFStream implements WebDataStream<Graph> {
    protected List<Consumer<Graph>> cs = new ArrayList<>();
    @Override
    public void addConsumer(Consumer<Graph> consumer) {
        cs.add(consumer);}
    @Override
    public void put(Graph g,long ts) {
        cs.forEach(c -> c.notify(g, ts)); }
}
```

Listing 3.5 RSP4J's WebDataStream implementation

Fig. 3.11 Streams interface

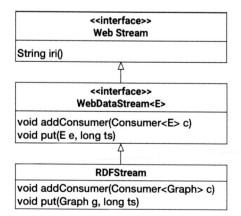

3.4.3 Operators

RSP4J's core includes separate interfaces for all the RSP-QL families of operators: *StreamToRelation*, *RelationToRelation*, and *RelationToStream*. These abstractions act both as lower level APIs for RSP4J's users and as a suitable entry point for extensions and optimizations. Moreover, each operator life cycle could be monitored independently. The **Stream-to-Relation** operator family bridges the world of RDF Streams to the world of finite RDF Data. RSP-QL defines a *Time-Based Sliding Window* operator for processing RDF Streams. When applied to an RDF stream, RSP-QL's S2R operator returns a function called Time-Varying Graph, that, given a time instant *t*, materializes an Instantaneous (finite) RDF Graph.

To correctly represent such behavior, RSP4J includes two interfaces, i.e., the `StreamToRelationFactory` interface and the `StreamToRelation` (S2R) operator. The former, exemplified in Listing 3.6, is used to instantiate the latter. It exposes the `build` method that takes the size of the window (width), the slide of the window (slide), and the start time (t0).

```
public interface StreamToRelationOperatorFactory<I,W> {
   StreamToRelationOp<I, W> build(long width, long slide, long t0);
}
```

Listing 3.6 RSP4J S2R operator factory interface

Listing 3.7 shows part of an implementation of the `StreamToRelationOperator Factory` that instantiates C-SPARQL's *Time-Based Sliding Window*.

```
public class CSPARQLS2RFactory implements S2RFactory<Graph,Graph> {
    private final Time time;
    private final Tick tick;
    private final Report report;
    private final ReportGrain grain;
    private ContentFactory<Graph, Graph> cf;

    @Override
    public S2R<Graph, Graph> build(long a, long b, long t0) {
            return new CSPARQLS2R( a, b, time, tick, report, grain,
    cf);

    }
}
```

Listing 3.7 The CSPARQL's time-based window operator in YASPER

The `StreamToRelation` operator is responsible for applying the windowing algorithm. In RSP4J, it is a special kind of `Consumer` that receives the data from the streams. Figure 3.12 shows the UML class diagram of the S2R Package. `CSPARQLWindowOperator` is a `StreamToRelation` operator, which creates a `TimeVarying<Graph>` if applied to a `WebDataStream<Graph>`.

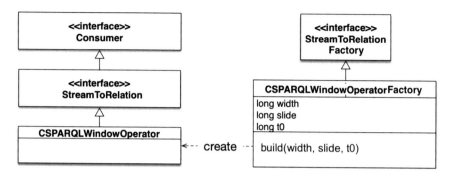

Fig. 3.12 UML class diagram for the S2R package as used by C-SPARQL

In RSP-QL, the **RelationToRelation** operator family corresponds to SPARQL 1.1 algebraic expressions evaluated over a given time instant. The evaluation of the Basic Graph Pattern produces a time-varying sequence of solution mappings, which can be consumed by SPARQL 1.1 operators. Listing 3.8 shows the RSP4J interface that covers this functionality. Similarly to the S2R operators, the interface is generic to let the RSP4J's users decide the internal representation of the query solution, e.g., the bindings.

```
public interface RelationToRelationOperator<W, R> {
       Stream<R> eval(Stream<W> sds);
    TimeVarying<Collection<R>> apply(SDS<W> sds);
}
```

Listing 3.8 RSP4J R2R operator interface

The **RelationToStream** operator family allows going back from the world of Solution Mappings to RDF Streams. According to RSP-QL, the evaluation of an R2S operator takes as input a sequence of time-varying solution mappings. In RSP4J, this idea is generalized as shown in Listing 3.9, i.e., RSP4J allows the user to also provide the solution mapping incrementally as soon as they are produced.

```
public interface RelationToStreamOperator<R, O> {
  O transform(R sm, long ts);
  Stream<O> eval(Stream<R> sml, long ts);
  Collection<O> eval(TimeVarying<Collection<R>> sml, long ts) ;
}
```

Listing 3.9 RSP4J R2S operator interface

3.4.4   SDS and Time-Varying Graphs

Like in SPARQL, the query specification and the SDS creation are closely related.
An RSP-QL dataset SDS is an extension of the SPARQL dataset to support the
continuous semantics. As indicated in Sect. 3.2.1, the SDS is time-dependent as
it contains time-varying graphs. RSP4J includes both the abstractions, i.e., the
SDS and the TimeVarying Graphs. Listing 3.10 shows RSP4J's SDS interface.
The generic parameter is inherited by the generic nature of RSP4J's Time-Varying
objects. The consolidate method consolidates the SDS content by recursively
consolidating every Time-Varying Object it contains.

```
public interface SDS<E> {
    default void consolidate(long t) {
        asTimeVaryingEs().forEach(tvg -> tvg.consolidate(ts));
    } ...
}
```

Listing 3.10 RSP4J's SDS interface

Listing 3.11 shows a Time-Varying Graph that is the result of the application of the Window Operator to an RDF Stream. The method materialize
consolidates the content at a given time instant *ts*. To this extent, it exploits
the StreamToRelationOperator interfaces, freezing and polling the active
window content. The *coalesce* method ensures only one graph, among those selected
during the windowing operation, is returned. According to RSP-QL, such graph
corresponds to the union of the RDF graphs in the window.

As time progresses, the *SDS* is reactively consolidated into a set of (named)
Instantaneous Graphs[5] at the time *t* at which a Time-Varying Graph is updated.
Thus, RSP4J includes the SDSManager and SDSConfiguration interfaces.
The former controls the creations, detection, and the interactions with the SDS; ide-
ally this represents a starting point for federated query answering and/or multi-query
optimization. Moreover, IT parametrizes the execution e.g., allowing alternative
windowing methods like Frames or Sessions, and output serializations like JSON-
LD or Turtle.

```
public class TimeVaryingGraph implements TimeVarying<Graph> {
    private IRI name;
    private StreamToRelationOperator<Graph, Graph> op;
    @Override
    public void materialize(long ts) {
        graph = op.getContent(ts).coalesce(); }...
}
```

Listing 3.11 YASPER's time-varying graph implementation

---

[5] Slowly evolving RDF graphs are represented as a (named) time-varying graph too.

3.4.5   Engine and Query Execution

This module includes the abstractions to control and monitor the engine and the query life cycle.

The **Engine** interface allows controlling RSP4J's capabilities, e.g., query registration and cancellation. It is based on the VoCaLS service feature idea [27]. Each engine can implement different interfaces, each of which corresponds to a particular feature. By querying the implemented interfaces, it is possible to list all the features exposed by the engine of choice, e.g., stream registration, RSP-QL support, or formatting the results in JSON-LD format.

```
public interface ContinuousQueryExecution<I, W, R, O> {
    DataStream<O> outstream();
    TimeVarying<Collection<R>> output();
    ContinuousQuery query();
    SDS<W> sds();
    StreamToRelationOp<I, W>[] s2rs();
    RelationToRelationOperator<W, R> r2r();
    RelationToStreamOperator<R, O> r2s();
    void add(StreamToRelationOp<I, W> op);
    Stream<R> eval(Long now);
}
```

Listing 3.12 RPS4J's continuous query execution interface

RSP4J core includes RSP-QL's reporting policies, i.e., On-Content-Change, Non-Empty-Content, Periodic, and On-Window-Close. `Report` is represented as a collection of `ReportingStrategies`. The `Content` interface represents the data items in the active window. It is generic, and by exposing the *coalesce*, it allows alternative implementations of the Time-Varying Graph functions.

The **ContinuousQueryExecution** interface, whose example is in Listing 3.12, represents the ever-lasting computation required by continuous queries. It allows monitoring and controlling the query life cycle. Moreover, in order to make the SDS and the operators involved in querying observable, the interface includes getters.

! The Use of Generics

In many of RSP4J's interfaces, generic types were used. This allows to extend the use of RSP4J beyond data models such as Graphs or Triples. RSP4J handles four types of parameters:

1. *I*: the generic type of the input stream
2. *W*: the type of the content of the window. Note that this is not always the same as *I*, e.g., the CQELS engine stores bindings instead of triples in its windows
3. *R*: the type the R2R operators operate on
4. *O*: the type of the output stream.

3.4.6 Composing Operators Using RSP4J's Operator API

To further facilitate the ease of prototyping, RSP4J provides an Operator API that allows to compose RPS engines using different implementations of RSP-QL operators. The Operator API creates *ContinuousPrograms*[6] based on a number of *Task*s. Each *Task* can be composed of a variety of S2R, R2R, and R2S operators. Listing 3.13 gives an example of the Operator API that composes an RSP engine based on different operators, all implementing the RSP4J interfaces. The *TaskBuilder* allows to add S2R operators using the *addS2R* method. This method accepts the name of the stream to operate upon, an object that implements RSP4J's *StreamToRelationOperator* interface and the name of the result (such that it can be used by later operators, like the R2R operator). The *addR2R* method defines on which results of the S2R operators the implementation of *RelationToRelationOperator* needs to be executed. Finally, the *addR2S* method defines how the results need to be reported. The *build* method combines the definition of these operators together and creates a *Task* that can be registered to the *ContinuousProgramBuilder*. The latter is defined in combination with the input and output streams.

```
Task<Graph, Graph, Binding, Binding> t =
  new Task.TaskBuilder()
    .addS2R("stream1", s2r, "w1")
    .addR2R("w1", r2r)
    .addR2S("out", new Rstream<Binding, Binding>()).build();
ContinuousProgram<Graph, Graph, Binding, Binding> cp =
  new ContinuousProgram.ContinuousProgramBuilder()
    .in(inputStream)
    .addTask(t)
    .out(outStream).build();
```

Listing 3.13 An example of the operator API

The Operator API also allows to compose tasks based on RSP-QL queries directly. Listing 3.14 gives an example of an RSP-QL query that is parsed straight onto a *Task* and used to build a *ContinuousProgram*. This method gives less flexibility, as the implementation of the operators is defined within the parsing process. When full flexibility is needed, the *TaskBuilder* is the best option.

---

[6] The *ContinuousProgram* and the implementation of the *ContinuousQueryExecution* interface.

```
ContinuousQuery<Graph, Graph, Binding, Binding> query =
  TPQueryFactory.parse(
    + "REGISTER RSTREAM :outStream AS "
    + "SELECT ?green "
    + "FROM NAMED WINDOW :win ON :colorStream [RANGE PT2S STEP PT2S]
      "
    + "WHERE {"
    + "   WINDOW :win { ?green a :Green .}"
    + "}");

Task<Graph, Graph, Binding, Binding> t =
  new QueryTask.QueryTaskBuilder()
    .fromQuery(query).build();
ContinuousProgram<Graph, Graph, Binding, Binding> cp =
  new ContinuousProgram.ContinuousProgramBuilder()
    .in(inputStream)
    .addTask(t)
    .out(query.getOutputStream()).build();
```

Listing 3.14 Parsing an RSP-QL task using the operator API

The goal of the Operator API is to ease the composition of custom RSP engines and to aid in the development of new operators. The latter is possible as RSP4J's YASPER implementation provides default implementations for all the operators.

? What Are the Specific Generic Types?

As you might have noticed, the above examples use four generic types, in this case *<Graph, Graph, Binding, Binding>*. These represent the instantiations of the *ContinuousQuery*, *Task*, and *ContinuousProgram* generic types *<I, W, R, O>*. As for getting to explain each of these types in details:

1. *Graph*: fills in the *I* generic type and is thus the type of the input stream
2. *Graph*: the second *Graph* denotes that *Graphs* are contained in the content of the window
3. *Binding*: the R2R operators operate on *Binding* objects
4. *Binding*: the second *Binding* type states that the output stream will contain *Binding* types as well.

3.4.7 Reasoning in RSP4J

Below, an example of how to integrate the C-Sprite algorithm in RSP4J is given. Building on the RSP4J's extensibility, it is easy to plug in the C-Sprite algorithm using the Operator API.

Listing 3.15 shows the Operator API code that combines the C-Sprite algorithm with the evaluation of a simple Triple Pattern. As the C-Sprite algorithm operates on a hierarchy of concepts, first on line 9 a hierarchy is created, which is made by a few simple definitions of *subClassOf* declaration. Next to the hierarchy, C-Sprite uses a Triple Pattern or Basic Graph Pattern to prune the hierarchy according to the types that are in the query, exclusively. Line 11 creates a simple Triple pattern that looks for all Warm colors in the hierarchy (?c a :WarmColor). On line 13, the creation of the C-Sprite algorithm as an R2R operator is seen, which takes the hierarchy and Triple Pattern as arguments. The generic types of the C-Sprite R2R operator are both *Graph*s, meaning that C-Sprite takes a *Graph* as input and outputs another *Graph*. The output *Graph* is the result of performing the hierarchical reasoning. On line 15 is the creation of an *R2RPipe*, which allows to combine (or pipeline) different R2R operators. The *R2RPipe* checks if the intermediate types of the various R2R operators match. In this case, the output of the C-Sprite operator is represented by *Graph*s, and the input of the Triple Pattern operator are *Graph*s as well. Note that the *R2RPipe* allows to combine the R2R operators, similarly as it has been seen in theoretical part of C-Sprite, in Sect. 3.3.2.1. Now it is possible to register this *R2RPipe* as R2R operator to the abstraction task on line 22.

```
// define the generator and input stream
StreamGenerator generator = new StreamGenerator();
DataStream<Graph> inputStream = generator.getStream("http://test/
    stream");
// define output stream
BindingStream outStream = new BindingStream("out");
//create the window operator
StreamToRelationOp<Graph, Graph> windowOperator =
    createWindowOperator(2000, 2000, "w1");
// create a simple hierarchy
HierarchySchema hierarchySchema = getHierarchySchema();
// create a simple triple pattern
RelationToRelationOperator<Graph, Binding> tp = createTriplePattern
    ();
// create the CSprite R2R
RelationToRelationOperator<Graph, Graph> cSpriteR2R = new CSpriteR2R
    (tp, hierarchySchema);
// create a R2R pipeline to combine CSprite and the TP evaluation
R2RPipe<Graph,Binding> r2rPipe = new R2RPipe<>(cSpriteR2R,tp);


// define the task and CP
TaskOperatorAPIImpl<Graph, Graph, Binding, Binding> t =
new TaskOperatorAPIImpl.TaskBuilder()
.addS2R("http://test/stream", windowOperator, "w1")
.addR2R("w1", r2rPipe)
.addR2S("out", new Rstream<Binding, Binding>())
.build();
ContinuousProgram<Graph, Graph, Binding, Binding> cp =
new ContinuousProgram.ContinuousProgramBuilder()
.in(inputStream)
.addTask(t)
.out(outStream)
.build();
```

Listing 3.15 Example of using CSprite in RSP4J through the operator API

Listing 3.16 shows part of the implementation of the C-Sprite R2R operator, more specifically the *eval* method, which is called to evaluate the operator. The operator is rather simple: for each triple in the received graphs, it performs C-Sprite's HierarchyMatch algorithm from Algorithm 2.

```
public Stream<Graph> eval(Stream<Graph> tvg) {
  return upwardExtendedGraph = tvg.map(
    g -> g.stream()
      .map(triple -> performHierarchyMatch(triple))
      .flatMap(Collection::stream)
      .collect(TripleCollector.toGraph())
  );
}
```

Listing 3.16 C-Sprite evaluation function: extends the graph stream and evaluates the R2R operator

With the simple use of the *R2RPipe*, it is possible to easily extend the behavior of the RSP engine. Note that the full code of the example is available on GitHub.[7]

3.5 Chapter Summary

In this chapter, we saw how to process data streams on the Web through RDF Stream Processing. We discussed the formal semantics of RSP engines by converging from CQL to RSP-QL. Next, we have discussed how the RSP-QL query language extends SPARQL to enable the processing of RDF Streams from a query perspective. In terms of reasoning over Web streams, we have shown that both forward and backward chaining impose serious problems. We have introduced C-Sprite as a hybrid approach to enable efficient reasoning over Web Streams. We have also introduced RSP4J, an API for the development of RSP engines under RSP-QL semantics. Through RSP4J's flexible model, we have shown how to query, define the various operators manually, or include reasoning when processing Web Streams.

References

1. Abiteboul, Serge, Richard Hull, and Victor Vianu. 1995. *Foundations of Databases*. Vol. 8. Boston: Addison-Wesley Reading.
2. Akidau, Tyler, Robert Bradshaw, Craig Chambers, Slava Chernyak, Rafael J. Fernández-Moctezuma, Reuven Lax, Sam McVeety, Daniel Mills, Frances Perry, Eric Schmidt, and Sam Whittle. 2015. The dataflow model: A practical approach to balancing correctness, latency,

---

[7] https://github.com/StreamingLinkedData/Book.

and cost in massive-scale, unbounded, out-of-order data processing. *Proceedings of the VLDB Endowment* 8: 1792–1803.

3. Al-Ajlan, Ajlan. 2015. The comparison between forward and backward chaining. *International Journal of Machine Learning and Computing* 5 (2): 106.

4. Arasu, Arvind, Shivnath Babu, and Jennifer Widom. 2006. The CQL continuous query language: Semantic foundations and query execution. *The VLDB Journal* 15 (2), 121–142.

5. Barbieri, Davide Francesco, Daniele Braga, Stefano Ceri, Emanuele Della Valle, and Michael Grossniklaus. 2010. Querying RDF streams with C-SPARQL. *SIGMOD Record* 39 (1): 20–26.

6. Bonte, Pieter, Riccardo Tommasini, Filip De Turck, Femke Ongenae, and Emanuele Della Valle. 2019. C-sprite: Efficient hierarchical reasoning for rapid RDF stream processing. In *DEBS*, 103–114. New York: ACM.

7. Calbimonte, Jean-Paul , Hoyoung Jeung, Oscar Corcho, and Karl Aberer. 2012. Enabling query technologies for the semantic sensor web. *International Journal On Semantic Web and Information Systems (IJSWIS)* 8 (1): 43–63.

8. Ceri, Stefano, and Jennifer Widom. 1991. Deriving production rules for incremental view maintenance. In *VLDB '91: Proceedings of the 17th International Conference on Very Large Data Bases*.

9. Dell'Aglio, Daniele, and Emanuele Della Valle. 2014. Incremental reasoning on RDF streams. In *Linked Data Management*, 413–435. Boca Raton: Chapman and Hall/CRC.

10. Dell'Aglio, Daniele, Emanuele Della Valle, Jean-Paul Calbimonte, and Óscar Corcho. 2014. RSP-QL semantics: A unifying query model to explain heterogeneity of RDF stream processing systems. *International Journal on Semantic Web and Information Systems (IJSWIS)* 10 (4): 17–44.

11. Dell'Aglio, Daniele, Emanuele Della Valle, Frank van Harmelen, and Abraham Bernstein. 2017. Stream reasoning: A survey and outlook. *Data Science* 1 (1–2): 59–83.

12. Falzone, Emanuele, Riccardo Tommasini, and Emanuele Della Valle. 2020. Stream reasoning: From theory to practice. In *Reasoning Web International Summer School* . Vol. 12258, 85–108. Berlin: Springer.

13. Gupta, Ashish, Inderpal Singh Mumick, and Venkatramanan Siva Subrahmanian. 1993. Maintaining views incrementally. *ACM SIGMOD Record* 22 (2): 157–166.

14. Hitzler, Pascal, and Krzysztof Janowicz. 2013. Linked data, big data, and the 4th paradigm. *Semantic Web* 4 (3): 233–235.

15. Isah, Haruna, Tariq Abughofa, Sazia Mahfuz, Dharmitha Ajerla, Farhana Zulkernine, and Shahzad Khan. 2019. A survey of distributed data stream processing frameworks. *IEEE Access* 7: 154300–154316.

16. Le-Phuoc, Danh, Minh Dao-Tran, Josiane Xavier Parreira, and Manfred Hauswirth. 2011. A native and adaptive approach for unified processing of linked streams and linked data. In *International Semantic Web Conference*, 370–388. Berlin: Springer.

17. Motik, Boris, Bernardo Cuenca Grau, Ian Horrocks, Zhe Wu, Achille Fokoue, Carsten Lutz, et al. 2009. OWL 2 web ontology language profiles. *W3C Recommendation* 27 (61). https://www.w3.org/TR/owl2-profiles/

18. Motik, Boris, Yavor Nenov, Robert Piro, and Ian Horrocks. 2019. Maintenance of datalog materialisations revisited. *Artificial Intelligence* 269: 76–136.

19. Motik, Boris, Yavor Nenov, Robert Edgar Felix Piro, and Ian Horrocks. 2015. Incremental update of datalog materialisation: The backward/forward algorithm. In *Twenty-Ninth AAAI Conference on Artificial Intelligence*.

20. Nenov, Yavor, Robert Piro, Boris Motik, Ian Horrocks, Zhe Wu, and Jay Banerjee. 2015. Rdfox: A highly-scalable RDF store. In *International Semantic Web Conference (2)*. Vol. 9367. *Lecture Notes in Computer Science*, 3–20. Berlin: Springer.

21. Oussous, Ahmed, Fatima-Zahra Benjelloun, Ayoub Ait Lahcen, and Samir Belfkih. 2018. Big data technologies: A survey. *Journal of King Saud University-Computer and Information Sciences* 30 (4): 431–448.

22. Ren, Xiangnan, and Olivier Curé. 2017. Strider: A hybrid adaptive distributed rdf stream processing engine. In *International Semantic Web Conference*, 559–576. Berlin: Springer.

23. Russell, Stuart J., and Peter Norvig. 2016. *Artificial Intelligence: A Modern Approach.* Malaysia: Pearson Education Limited.
24. Sagiroglu, Seref, and Duygu Sinanc. 2013. Big data: A review. In *2013 International Conference on Collaboration Technologies and Systems (CTS)*, 42–47. Piscataway: IEEE.
25. Staudt, Martin, and Matthias Jarke. 1995. *Incremental Maintenance of Externally Materialized Views.* Citeseer.
26. Terry, Douglas B., David Goldberg, David A. Nichols, and Brian M. Oki. 1992. Continuous queries over append-only databases. In *Proceedings of the 1992 ACM SIGMOD International Conference on Management of Data, San Diego, California, USA, June 2–5, 1992*, 321–330. New York: ACM Press.
27. Tommasini, Riccardo, Yehia Abo Sedira, Daniele Dell'Aglio, Marco Balduini, Muhammad Intizar Ali, Danh Le Phuoc, Emanuele Della Valle, and Jean-Paul Calbimonte. 2018. Vocals: Vocabulary and catalog of linked streams. In *International Semantic Web Conference (2)*. Vol. 11137. *Lecture Notes in Computer Science*, 256–272. Berlin: Springer.
28. Tommasini, Riccardo, Pieter Bonte, Femke Ongenae, and Emanuele Della Valle. 2021. RSP4J: An API for RDF stream processing. In *The Semantic Web - 18th International Conference, ESWC 2021, Virtual Event, June 6-10, 2021, Proceedings*, ed. Ruben Verborgh, Katja Hose, Heiko Paulheim, Pierre-Antoine Champin, Maria Maleshkova, Óscar Corcho, Petar Ristoski, and Mehwish Alam. Vol. 12731. *Lecture Notes in Computer Science*, 565–581. Berlin: Springer.
29. Valle, Emanuele Della, Stefano Ceri, Frank van Harmelen, and Dieter Fensel. 2009. It's a streaming world! reasoning upon rapidly changing information. *IEEE Intelligent Systems* 24 (6): 83–89.

Chapter 4
Streaming Linked Data Life Cycle

Abstract The term Linked Data, which was invented by Sir Tim Berners-Lee, refers to a set of best practices for publishing data using Semantic Web technologies. Although Linked Data can be dynamic, the growing availability of data streams poses new challenges that require extensions to both the best practices and the technologies. This chapter presents a proposal for a Streaming Linked Data life cycle. In particular, it focuses on the data creation process developing seven steps and aims at making data streams findable, accessible, reusable, and interoperable. Such steps are *Name*, which focuses on the identification of data streams as Web resources by the assignment of an IRI; *Model*, which requires knowledge representation skills to represent both the data stream metadata and the data themselves, by taking into account their ephemeral nature; *Shape*, which discusses the method of the data model of choice, i.e., how to represent the minimal unit of information. Indeed, RDF leads to different design decisions that impact the processing performance; *Annotate*, which focuses on the transformation of raw streaming data into RDF streams. This optional step clarifies the need for interoperable data format for data streams, too; *Describe* highlights the need for extensive and interoperable metadata that aid the stream discovery; *Serve* focuses on the format, protocol, and services that are responsible of the data sharing in practice; *Process* concludes the life cycle indicating the remainder steps, summarized later on in this chapter. In addition to the steps definition, this chapter introduces relevant resources and guides the readers through their practice with practical examples.

4.1 Introduction

This chapter presents a proposal for a Streaming Linked Data life cycle focusing on the creation process. [32]. The presented life cycle, consisting of seven steps, aims at making data streams findable, accessible, reusable, and interoperable [37]. In practice, this chapter highlights the community achievements related to pursuing such goal; it introduces important resources that can aid practitioners in the publishing process, e.g., ontologies [13, 34] and systems [22, 33], and describes how to use them in practice with reference to the seven steps.

Fig. 4.1 Publication starting point [32]

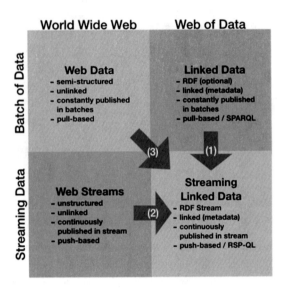

Tommasini et al. [32] identified three situations a practitioner might find when dealing with publishing Streaming Linked Data. Those cases are depicted in Fig. 4.1. The ultimate goal is identified by the *lower-right* quadrant, while the other quadrants present possible starting points, i.e., (upper-left) Web Data published in batches; (upper-right) Linked Data published in batches; and (lower-left) Web Data published as streams.

The remainder of this chapter is organized as follows: Sect. 4.2 introduces preliminary concepts. Section 4.3 presents the streaming linked data life cycle, discussing in detail all the life cycle steps. Finally, Sect. 4.4 concludes this chapter summarizing the essential points.

4.2 Linked and FAIR Data

The term Linked Data refers to a set of best practices for publishing data using Semantic Web technologies [6]. Tim Berners-Lee invented the term to highlight the value of making connections across datasets available on the World Wide Web. Several life cycles for Linked Data have been proposed in the literature with the intent of distilling the most important steps that concern linked data publications [25]. However, the simplicity of the idea can be summarized in four principles, i.e.,

1. Use URIs as names for dereferenced things
2. Use HTTP URIs so that people can look up those names
3. When someone looks up a URI, provide useful information using standards
4. Include links to other URIs, so that they can discover more things.

Stage		Explanation	Example of data format
★	1st stage	Available on the web (whatever format) but with an open license, to be Open Data	PDF, JPG
★★	2nd stage	Available as machine-readable structured data (e.g. excel instead of image scan of a table)	XLS, DOC
★★★	3rd stage	as (2) plus non-proprietary format (e.g. CSV instead of excel)	XML, CSV
★★★★	4th stage	All the above, plus: Use open standards from W3C (RDF and SPARQL) to identify things, so that people can point at your stuff	RDF
★★★★★	5th stage	All the above, plus: Link your data to other people's data to provide context	Linked-RDF

Fig. 4.2 Five stars of linked data

These four principles guide the Web of Data toward a global data space. Coherently with the Web design principles, Linked Data principles lay the foundations for extending the Web architecture. Figure 4.2 shows how, by following the aforementioned principles, it is possible to publish high-quality data. The fourth and fifth stages are those of main interest. Data shared using open and machine-readable formats, available and with links to other resources, are the ultimate goal.

To ease data discovery, public data catalogs like Google's Dataset Search are emerging, guided by the FAIR principles.[1] The initiative's pillars, i.e., Findable, Accessible, Interoperable, Reusable, aim at improving the way (scientific) data are shared and foster interoperability. Therefore, they act as requirements for applications focused on data sharing. While presenting streaming linked data life cycle, it is important to take the FAIR principles into account. This section summarizes the FAIR points to guide the reader within the chapter.

Findable (F1) Data should be assigned unique and persistent identifiers, e.g., DOI or URIs. (F2) Data should be assigned metadata that includes descriptive information, data quality, and context. (F3) Metadata should explicitly name the persistent identifier since they often come in a separate file. (F4) Identifiers and metadata should be indexed or searchable.

Accessible (A1) Data and metadata should be accessible via (a) free, (b) open-sourced, and (c) standard communication protocols, e.g., HTTP or FTP. Nonetheless, authorization and authentication are possible. (A2) Metadata should be accessible even when data are no longer available.

Interoperable (I1) Data and metadata must be written using formal languages and shared vocabularies that are accessible to a broad audience. (I2) Such vocabularies should also fulfill FAIR principles. (I3) Data and metadata should use qualified references to other (meta)data.

---

[1] http://go-fair.org.

Reusable (R1) Data should adopt an explicit license for access and usage. (R2) Data provenance should be documented and accessible. (R3) Data and metadata should comply with community standards.

4.3 The Steps of the Life Cycle

This section presents a proposal for a streaming linked data life cycle based on the one introduced in [32]. In particular, the following steps constitute the life cycle, which Fig. 4.3 portrays:

- *Name*, which focuses on the identification of data streams as Web Resources by the assignment of an IRI.
- *Model*, which requires knowledge representation skills to represent both the data stream metadata and the data themselves, by taking into account their ephemeral nature.
- *Shape*, which discusses the method of the data model of choice, i.e., how to represent the minimal unit of information. Indeed, RDF leads to different design decisions that impact the processing performance.
- *Annotate*, which focuses on the transformation of raw streaming data into RDF streams. This optional step clarifies the need for interoperable data format for data streams, too.
- *Describe* highlights the need for extensive and interoperable metadata that aid the stream discovery.
- *Serve* focuses on the format, protocol, and services that are responsible of the data sharing in practice.
- *Process* concludes the life cycle indicating the remainder steps, summarized later on in this chapter.

The following sections will go more into details for each given step. They will present state-of-the-art solutions for each one of them, briefly compare them, and provide an exhaustive solution that reflects the authors' view.

Fig. 4.3 Streaming linked data life cycle

4.3.1 Identify

The identification step of the life cycle aims at distinguishing relevant resources and designing IRIs that identify them. Data streams, like datasets, represent a collection of data points. Thus, two kinds of resources are critical when discussing streaming data on the Web, i.e., the stream itself and the ones the stream contains. Tommasini et al. [32] highlight some important aspects of the data stream, as they impact how data should be managed and identified. That is, they are unbounded and ordered [30]:

- *Unboundedness* relates to the impossibility of storing the stream as whole
- *Order* impacts how data are consumed, i.e., sequentially as soon as they arrive.

Definition 4.1 A Web data stream is a Web Resource that identifies an unbounded ordered collection of pairs (o, i), where o is a Web resource, e.g., a named graph, and i is a metadatum that can be used to establish an ordering relation, e.g., a timestamp.

Definition 4.1 from [32] helps characterizing stream resources. In particular, the definition makes no assumption on the data that will be received. Instead, it decouples the identification problem between the data stream and the resource it contains. Linked Data best practices for good IRI design prescribe the usage of HTTP IRIs to identify Web resources. While such best practices apply also to the Web Stream resource, it is important to observe that data access does not work in the same way. Indeed, accessing streaming data requires a continuously open connection (e.g., WebSocket). Hereafter, the section introduces two approaches to solve identification, respectively, from Sequeda and Corcho and from Barbieri and Della Valle.

Sequeda and Corcho proposed three innovative IRI schemes to identify sensors and their observations [28]. In particular, the URI scheme below identifies a window (start, end):

```
http://linkeddata.stream/sensor/name/%start time%,
                    %end time%
```

Listing 4.1 shows an example from [28] that identifies a sensor as a streaming data source and an observation as a streaming element. Sequeda and Corcho's vision actually goes beyond streaming data and incorporates spatiotemporal metadata. Indeed, sensor data are often associated with location metadata.[2]

---

[2] The interested reader can consult the book [20].

```
hrs:1 a s:Sensor   ;
    s:measures [
        _measurement a hr:HeartRateMonitor
    ].

hrs:1 s:measures hrs:1/2022-07-15 17:00:00 .
    rdf:type
    hr:HeartRateMonitor;
    hr:heartRate "74";
    hr:timestamp "2022-07-15 17:00:00"^^xsd:dateTime

hrs:3 s:measures hrs:1/2022-07-15 17:05:00 .
    rdf:type hr:HeartRateMonitor;
    hr:heartRate "58";
    hr:timestamp "2022-07-15 17:05:00"^^xsd:dateTime
```

Listing 4.1 Sensor and Observation [28].

```
:sgraph1 sld:lastUpdate "τ_{i+1}"^^xsd:dataTime ;
    sld:expires "τ_{i+2}"^^xsd:dataTime ;
    sld:windowType sld:logicalTumbling ;
    sld:windowSize "PT1H"^^xsd:duration .

:igraph1 :receivedAt "τ_i"^^xsd:dataTime ;
    :rdfs:seeAlso :sgraph1.
:igraph2 :receivedAt "τ_{i+1}"^^xsd:dataTime ;
    :rdfs:seeAlso :sgraph1.

:igraph1 { # I-Graph at τ_i in TriX Syntax
    :broker1 :does [ :tr1 :with "$ 1000" ]
    }
:igraph2 { # I-Graph at τ_{i+1} in TriX Syntax
    :broker1 :does [ :tr2 :with "$ 3000" ] .
    :broker2 :does [ :tr3 :with "$ 2000" ] .
    }
```

Listing 4.2 sGraph and iGraphs [5].

Barbieri and Della Valle [5] recommended to identify streams using IRIs that resolve a named graph containing all the relevant metadata (called sGraph) and IRIs to identify the single element in the stream (called iGraphs) via timestamping. iGraphs and sGraph link to each other using the property *rdfs:seeAlso*, while the property *:receivedAt* attaches a timestamp to the iGraphs using *xsd* literals. For example, the sGraph in Listing 4.2 lines 1–4 is the result of resolving http://stockex.org/transactions. While the iGraphs in Listing 4.2 6–8 and 12–19 are the result of resolving http://stockex.org/transactions/urlenconde(τ_i) and http://stockex.org/transactions/urlenconde(τ_{i+1}). Thus, Barbieri and Della Valle propose two URI schemes:

```
http://linkeddata.stream/%stream-name%
http://linkeddata.stream/%stream-name%/urlenconde(%timestamp%)
```

In practice, unboundedness leads to the *ephemerality* of the stream elements. Indeed, clients accessing the stream will receive data sequentially starting from the moment they connect. If not explicitly stored, historical data will not be accessible.

To overcome this issue, Barbieri and Della Valle backlink iGraphs to the origin sGraph, which remains referentiable. Sequeda and Corcho suggest to encode such dependency within the IRI schema. However, extending the IRI schema may cause confusion because each resource becomes independent from the stream. Instead, Hash IRIs and fragment identifiers allow to bind the stream elements to the stream resource without needing an extra triple nor using a custom IRI scheme, i.e.,

```
http://linkeddata.stream/{stream-name}#{igraph-id}.
```

Use Hash IRIs

Using Hash IRIs with fragment identifiers ensures that the stream content representation is directly associated with the stream resource itself.

Example 4.1 (Cont'd) Carrying on the running example for the identification step, the color stream can be identified by the base URL http://linkeddata.stream. Moreover, the following URI schemas will be used to identify Web Resources that are relevant for the book:

1. http://linkeddata.stream/ontologies/color-ontology
2. http://linkeddata.stream/resource/color-stream
3. http://linkeddata.stream/resource/color-stream#color-obs-id.

4.3.2   Model

This step of the life cycle aims at describing the application domain from which data come. To this extent, ontologies and models are designed and implemented using formal languages like RDFS or OWL in order to capture the domain knowledge in machine-understandable way. At the time of writing, a stream-specific knowledge representation methodology does not exist. Nonetheless, several ontologies that are prominent in the stream reasoning literature, e.g., Frappe [2] and SAO [13], among many others, are available. In particular, Frappe and SAO were specifically designed for stream reasoning tasks. Therefore, this section studies SAO's and Frappe's conceptualizations and presents their peculiar characteristics. Moreover, it elicits general guidance for modeling Web streams starting from these two pioneering works.

The stream processing literature found a general agreement about the most relevant abstractions. In particular, streaming data are normally divided in *instantaneous* and *time-varying*. The former identifies data items that are valid at a specific point in time, e.g., elements in a data stream; the latter identifies data that change over time, e.g., the content of a time-based window. The data dichotomy was introduced by Arasu et al. [1] to the extent of formalizing relational continuous queries, and it was extended to RSP by Dell'Aglio et al. [9] to operate on RDF graphs. In practice, time-varying concepts are the results of applying a continuous *stateful* computation to instantaneous data. Although purely static data are not included in

Table 4.1 Concepts and
classes for SAO and Frappe

Concept	SAO	Frappe
Instantaneous	Point, StreamEvent	Event
Time-varying	StreamData, Segment	Pixel, Frame
Continuous	ColorSynthesis	Synthesize
Static	StreamAnalysis	Grid, Cell, and Place

the formalization, they are often used in practice. Indeed, the formalization can be extended to represent data that are not subject to changes [21]. The section shows how SAO and Frappe distinguish between instantaneous, time-varying, continuous, and static concepts. Moreover, it continues the running examples modeling the color domain around the same abstractions. Table 4.1 summarizes the findings of comparing the two ontologies with the stream processing abstract concepts.

Distinguish Instantaneous and Time-Varying Concepts

Instantaneous concepts are valid at a specific point in time, e.g., a sensor observation. On the other hand, time-varying concepts change over time. They represent the two conceptual sides of a streaming application that operate on the elements in a stream.

The **Stream Annotation Ontology (SAO)**[3] allows publishing derived data about IoT streams. The vocabulary (Fig. 4.4) allows to describe the aggregation transformations in depth. SAO relies on PROV-O to track the aggregation provenance and OWL-Time for the temporal annotations [18]. SAO adopts two classes to represent instantaneous data, i.e., sao:Point and sao:StreamEvent. The former represents a single data point in a stream of sensor observation (see Listing 4.3 at line 4.), while the latter expresses an artificial classification of the stream elements (see Listing 4.3 at line 10.). Moreover, SAO includes two Time-Varying concepts, i.e., sao:StreamData that represents data points results of an aggregation, and sao:Segments, which represent a portion of a stream. SAO includes the class sao:StreamAnalysis to represent a continuous computation (see Listing 4.3 at line 20). As mentioned, the result is a time-varying concept. In SAO, the conceptualization is limited to the class sao:Sensor that represents a streaming data source (see Listing 4.3 line 2).

Mind the Static-Streaming Gap

When modeling the application domain it is important to take into account links between static and streaming data. Such link should be time-agnostic and ease tasks like streaming data enrichment and augmentation.

---

[3] http://iot.ee.surrey.ac.uk/citypulse/ontologies/sao/sao.

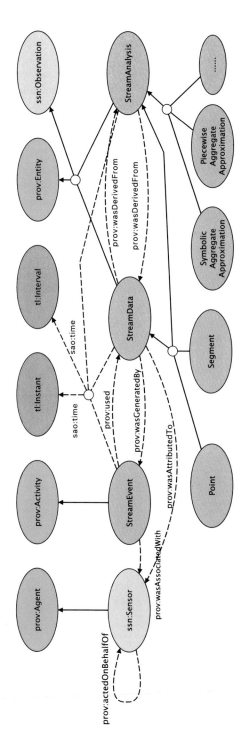

Fig. 4.4 SAO schema

Frappe[4] is an OWL 2 QL vocabulary for spatio-temporal data analytics. It borrows its conceptualization (Fig. 4.5) from the domain of photography [2]. It represents the world as a sequence of frames and events occurring within a spatio-temporal context (see Listing 4.4). Frappe borrows the instantaneous concept *Event* from the EventOntology, which is meant to represent something that happens in the real world at a given time. Notably, both SAO and Frappe represent and make use of OWL-Time (see Listing 4.4 at line 17). Similarly to SAO, Frappe identifies `frp:Pixel` and `frp:Frame`, which, respectively, correspond to the results of continuous computations over the Events occurring in a Cell or in a Grid (see Listing 4.4 at lines 26 and 29). Frappe represents continuous computations as `frp:Capture` and `frp:Synthesize` (see Listing 4.4 at line 34). In Frappe, the classes Grid, Cell, and Place, which represent the spatial context, are assumed to be static (see Listing 4.4 at line 2). Note that data are treated as static when they are not assumed to change during the query execution. Notably, static data are considered different concepts not subject to time-like units of measure.

Representing Continuous Computations

Identifying continuous computations allows to characterize the streaming application workflow. Moreover, they link instantaneous and time-varying concepts, enabling the provenance tracking for the streaming analyses and transformations.

```
<sensor>  a  ssn:Sensor ;
      ssn:observes  qoi:Freshness .

:sensorRec1 rdf:type sao:Point ;
    prov:wasAttributedTo <sensor> .

:sensorRec2 rdf:type sao:Point ;
    prov:wasAttributedTo <sensor> .

:traffic-sensor-recording-619 rdf:type sao:StreamEvent ;
    prov:wasAsscoatedWith :government ;
  prov:used [ rdf:type :sensorRec1, :sensorRec2 ] ;
  sao:time  [ rdf:type  tl:Interval ;
         tl:at "2014-02-13T08:25:00"^^xsd:dateTime ;
         tl:duration "PT15H30M"^^xsd:duration ] .

:freshness-traffic-619 rdf:type qoi:Freshness ;
  qoi:value "2014-02-13T08:25:00"^^xsd:dateTime .

:sax_AverageSpeedSample rdf:type  sao:SymbolicAggregateApproximation;
  rdfs:label "The sax representation of the traffic sensor recording obtained from Aarhus
      City.";
  sao:value "bbbbacdd";
  sao:alphabetsize "4"^^xsd:int ;
  sao:segmentsize "8"^^xsd:int ;
  prov:wasGeneratedBy :traffic-sensor-recording-619;
  qoi:hasQoI :freshness-traffic-619 .
```

Listing 4.3 SAO Example.

---

[4] http://lov.okfn.org/dataset/lov/vocabs/frapp.

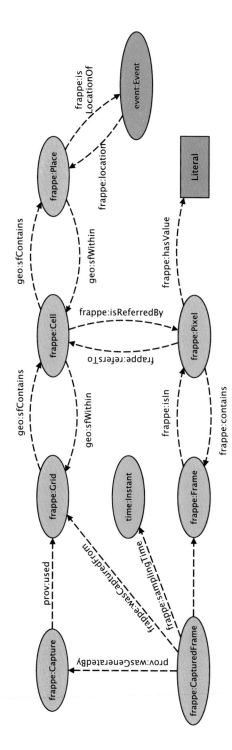

Fig. 4.5 Frappe schema

```
:Grid_1 gs:sfContains :Cell_1, :Cell_2 .
:Cell_1 a fr:Cell ;
  rdfs:label "39460"^^xsd:long ;
  fr:isReferredBy :1356995100000_39460 ;
  gs:sfContains :place1 ;
  gs:sfWithin :Grid_1 .
:place1 a fr:Point ;
  gs:asWKT "POINT( 40.715008 -73.96244 )"^^gs:wktLiteral ;
  gs:sfWithin :Cell_1 .

:E_A a :PickUpEvent ;    a event:Event ;
  event:time [ a time:Instant ;
    time:inXSDDateTime "2013-01-01T00:00:00"^^xsd:dateTime ];
  fr:location :place1 ;
  :hackLicense "E7750A37CAB07D0DFAF7E3573AC141"^^xsd:string;
  :medallion "07290D3599E7A0D6209346EFCC1FB5"^^xsd:string .
:E_B a :DropOffEvent ;  a event:Event ;
  event:time [ a time:Instant ;
    time:inXSDDateTime "2013-01-01T00:02:00"^^xsd:dateTime ];
  fr:location :B ; :connected :E_A ;
  :paymentType "CSH"^^xsd:string;
  :fareAmount "3.5"^^xsd:double;
  :totalAmount "4.5"^^xsd:double;
  :tripDistance "0.44"^^xsd:long; :tripTime "120"^^xsd:long.

:1356995100000_39460 a fr:Pixel ;
  fr:isIn :1356995100000 ;
  fr:refers :Cell_1 .
:1356995100000 a fr:CapturedFrame ;
  fr:contains :1356995100000_39460, :1356995100000_39461 ;
  fr:wasCapturedFrom :Grid_1 ;
  prov:wasGeneratedBy :1356995100000 ;
  fr:samplingTime [ a time:Instant ;
    time:inXSDDateTime "2013-01-01T00:05:00"^^xsd:dateTime] .
```

Listing 4.4 Frappe DEBS Example.

Example 4.2 (Cont'd) Carrying on with the running example about the color stream for the modeling step, the background knowledge about the application domain can be considered. Indeed, during this step, it is critical to identify relevant resources, collect data samples, and formulate information needs. Figures 4.6 and 4.7 describe colors according to their structure and sentiment, respectively. In particular, they help identifying (i) primary, secondary, and tertiary colors; (ii) cool vs. warm colors; (iii) the association between primary and secondary colors and emotions, e.g., red and anger. Figure 4.9 completes the example indicating how data are provisioned, i.e., colors are the result of a sensor observation that measures the light spectrum. The role of a knowledge engineer is to capture the model using a formal language that allows a machine-related understanding of such information. Figure 4.8 exemplifies a possible way to represent the domain knowledge about the colors' structure. Sentiments are represented as instances of a :Sentiment class and associated with the color classes. The colors taxonomy represents them as classes in order to provide different hierarchies based on their composition and "temperature." Thus, individuals are instances of the colors to be identified. Figure 4.9 helps mapping the sensor measurement into symbolic representations of colors, e.g.:

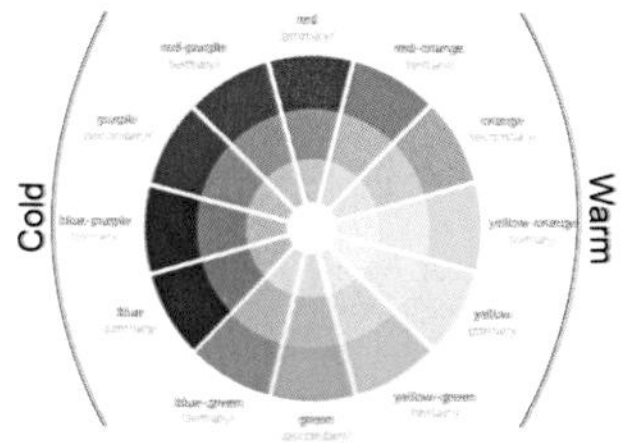

Fig. 4.6 Domain knowledge
about colors' structure

Fig. 4.7 Domain knowledge
about colors' sentiment

- if the light frequency is between 650 and 605, the light is perceived as *blue*;
- if the light frequency is between 605 and 545, the light is perceived as *green*;
- if the light frequency is between 545 and 480, the light is perceived as *yellow*;
- if the light frequency is between 480 and 380, the light is perceived as *red*.

The OWL 2 version of the color ontology is available in Listing 4.5 and online.[5]

As mentioned, it is important to distinguish between instantaneous and time-varying concepts. These concepts are essential to formulate information needs and present them as continuous transformations. Moreover, if any static data abstraction emerges, it should be included in the modeling. For what concerns the color perception, a generic sensor observation captures the idea of the instantaneous concept (Table 4.2).

Figure 4.9 tells what color it is possible to sense. The class :PerceivedColor as opposed to :DerivedColor identifies instantaneous concepts as well as time-varying ones resulting from the co-occurrence of multiple perceived colors. In the colored stream example, the acts of composing and decomposing colors are continuous, i.e., they are repeatedly evaluated over a stream of input, and their output is a stream as well. In particular, the result of color composition (:Synthesis) is a stream of derived colors. Finally, the association between colors and sentiments does not depend on any temporal annotation. Thus, sentiments are represented as static knowledge that can be joined with the stream of colors.

---

[5] https://linkeddata.stream/ontologies/colors.owl.

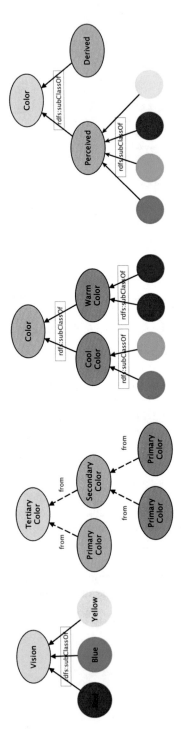

Fig. 4.8 Graphical representation of the color ontology

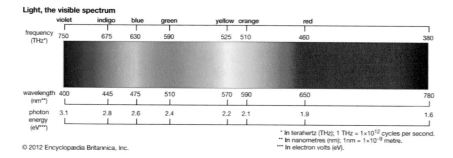

Fig. 4.9 Light spectrum overview

```
:Color a owl:Class .
:Sentiment a owl:Class .

:fear      a :Sentiment . :anger     a :Sentiment . :sadness   a :Sentiment .
:anxiety   a :Sentiment . :surprise  a :Sentiment . :happiness a :Sentiment .

:from a owl:ObjectProperty ; rdfs:domain :Derived ; rdfs:range :Color .
:indicates a owl:ObjectProperty ; rdfs:domain :Color ; rdfs:range :Sentiment .

:ColorSynthesis a owl:Class ;

:Primary   a owl:Class ; rdfs:subClassOf :Color .
:Secondary a owl:Class ; rdfs:subClassOf :Color .
:Tertiary  a owl:Class ; rdfs:subClassOf :Color .

:Cold a owl:Class ; rdfs:subClassOf :Color . :Warm a owl:Class ; rdfs:subClassOf :Color .

:Red    a owl:Class ; rdfs:subClassOf :Primary, :Perceived, :Warm ; :indicates :anger .
:Blue   a owl:Class ; rdfs:subClassOf :Primary, :Perceived, :Cold ; :indicates :sadness .
:Yellow a owl:Class ; rdfs:subClassOf :Primary, :Warm ;  :indicates :surprise .

:Green  a owl:Class ; rdfs:subClassOf :Secondary, :Perceived ;
   [ a owl:Restriction ; owl:onProperty :from ; owl:someValuesFrom :Yellow],
   [ a owl:Restriction ; owl:onProperty :from ; owl:someValuesFrom :Blue ] ; :indicates :
     happiness .
```

Listing 4.5 Color Ontology.

Table 4.2 Color: concepts and classes

Concept	Class
Instantaneous	PerceivedColor
Time-varying	DerivedColor
Continuous	Synthesis/decomposition
Static	Sentiment

> **Exercise**

Try define the Purple and the Orange color classes and link them to their respective sentiment with reference to Fig. 4.7. Are they perceived colors?

Fig. 4.10 Triple-based RDF
stream

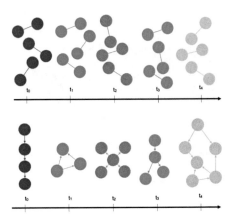

Fig. 4.11 Graph-based RDF
stream

4.3.3 Shape

The shaping step of the life cycle is about molding the stream data items to ease
their consumption and processing. The stream processing literature distinguishes
between *structured* and *semantic* data streams [10]. The former indicates a data
stream whose instantaneous items content is constrained by a schema. For example,
in relational stream processing, every element in a data stream must comply to
one and only one schema. The latter refers to a data stream whose instantaneous
elements can be interpreted with reference to a given model. In practice, the model is
often written in a formal language (e.g., RDFS or OWL), and the interpretation follows traditional deductive reasoning tasks. For structured streams, the schema is typically defined in advance. Thus, the shaping step reduces to either identify a subset
of the schema to publish or extend the schema by combining multiple data sources.
On the other hand, semantic streams typically adopt schema-less data model, e.g.,
RDF. Therefore, shaping comes down to deciding the minimal structural unit that
constitutes the stream data item. Shaping differs from modeling because at this stage
the conceptualization is already specialized in an ontological schema. Instead, at
which level of granularity the data are provisioned is yet to be defined. In order to
decide the final shape, the following important questions should be answered:

? Questions to Answer

- What is the minimal unit of information that is useful for the processing?
- What are the requirements in terms of throughput and latency?

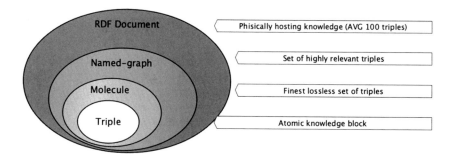

Fig. 4.12 RDF molecule [11]

In the stream reasoning state, two approaches are popular for shaping streaming data, i.e.,

- triple-based RDF streams, which use individually timestamped triples to punctuate the data stream
- graph-based RDF streams, which use timestamped (named) graphs to punctuate the data streams.

The two approaches have been shown to be equivalent from a formal point of view [9]. Indeed, it is always possible to build a graph-based stream from a triple-based one, grouping the elements by time. Vice versa, transforming a graph-based stream into a triple-based one requires a stateless operation that flattens the stream content (e.g., a flatMap).

However, the two models have a different impact on the processing performance. In general, triple-based streams (see Fig. 4.10) optimize for latency, as the ingestion proceeds alongside the input parsing (which happens node by node, triple by triple, and graph by graph). They impact redundancy as equivalent nodes are repeated multiple times. On the other hand, graph-based streams (see Figs. 4.10 and 4.11) optimize better for throughput since they are a de facto micro-batching mechanics, but they require a more sophisticated parsing. The finer-grained punctuation makes triple-based streams more subject to noise like late arrivals, while graph-based molecules have a higher risk of information loss. Finally, triple-based streams are quite generic and suitable for any processing task, including incremental maintenance (i.e., they should be further interpreted as additions and deletions). Instead, graph-based streams pave the road to query-driven stream shaping.

Definition 4.2 (RDF Molecule) Given an RDF graph G, an RDF molecule M ⊂ is the finest set of triples $\{t_1, t_2, ..t_n\}$ that decomposes the graph without loss of information [11].

Alternative approaches are possible: ERI Streams are a sequence of contiguous blocks of triple with the same subjects [12]. The approach then applies an encoding procedure similar to that of the Efficient XML Interchange (EXI) format adapted for RDF. Each group is multiplexed into lists of items with similar values, which are well suited for standard compression algorithms. ERI Stream adopts Ding et al.'s notion of RDF Molecule to determine groups [11]. In practice, the notion can help generalizing the idea of RDF stream shaping. According to Definition 4.2, RDF molecules are slightly more informative than individual RDF triple, yet they are not identified as named graphs (see Fig. 4.12). Notably, Ding's molecules are not identified a priori and they are meant for provenance tracking. Notably, an agreement between the stream molecule and the shape of continuous queries might allow performance optimization. Figure 4.11 shows a graph stream in which every graph follows one of the most common SPARQL query shapes (see and Fig. 4.11), which are also compatible with RSP-QL, that could inspire specific molecular structures.

Align Shape and Instantaneous Concept

The stream shape directly impacts different aspects of the ingesting, e.g., parsing and filtering. Aligning it with the instantaneous concepts simplifies the design of efficient processing steps.

4.3.4 Annotate

The annotation step of the life cycle is about converting streaming data into a machine-readable format. As for Linked Data, RDF is the data model of choice to publish data on the Web. However, streaming data are not produced directly in RDF. For instance, sensors usually push observations using a tabular format like CSV or document formats like JSON. Often, data are compressed to save bandwidth and reduce transfer costs. In all the above cases, a conversion mechanism must be set up to transform the data stream into machine-readable format. To design an adequate conversion pipeline, the following important questions should be answered:

? Questions to Answer

- What is the input data model and what is the expected output one?
- Is there any contextual information that can be used during the conversion?
- What mapping language is more suitable for the conversion?
- How do we treat the streaming temporal dimension during annotation?

The conversion pipeline should make use of the domain ontologies designed during the modeling step. Additionally, streaming data may be combined with contextual domain knowledge, capturing the domain information collected in an ontological model. The result of the annotation should shape the RDF stream according with the decision made at the previous step of the life cycle. Thus, the RDF molecule plays an important role also in the annotation process.

Use Semantic Streams

The annotation step is option, yet—like for Linked Data—the recommendation is to adopt semantic streams to foster data sharing and integration.

Technologies like R2RML are adequate to set up static data conversion pipelines, but they present some limitations when having to deal with data streams. In particular, due to the infinite nature of streaming data, the conversion mechanism does terminate if it operates under the same assumptions used for dataset annotation. In the literature, the most common workaround consists in taking into account the stream one element at a time. Alternatively, one can use a window-based Stream Processing engine to transform the input stream by means of a continuous query. The latter approach has been only described, yet never implemented. A notable exception is Morph$_{stream}$ [8], which extends the mapping language to include a windowing semantics, but rather than annotating the data, it uses the mapping to enable ontology-based streaming data access [17].

Avoid Time-Based Streaming Data Annotation

Due to the limitations of existing mapping languages, it is recommended to annotate element-by-element or adopt a micro-batching approach to ensure the termination of the conversion process.

Example 4.3 Carrying on the running example about the color stream, the annotation step should make use of the Color ontology in Listing 4.5. Figure 4.13 shows some observation made by an RGB sensor about the light spectrum. In practice, observation data are typically represented as JSON or CSV. For simplicity, we assume the latter.

To transform the CSV stream into an RDF Stream, the conversion pipeline must apply a transformation rule that can be specified using the RML mapping language. RML is preferred over the R2RML W3C recommendation because it supports processing CSV directly, as well as a number of other formats and input protocols. Listing 4.6 shows a subset of the mapping that focuses on the Blue color. Functions are used to discriminate between the frequencies, while several similar rules are deployed for the colors specified in the ontology. Notably, the mapping uses the more readable YARRML syntax. Finally, Fig. 4.14 displays the resulting RDF stream denoting sample occurrences of Blue. At the time of writing, the conversion pipeline was implemented using CARML[6] or RMLStreamer,[7] which support streaming-like annotation given a standard way for iterating over the input data, i.e., line-by-line for CSV or document-by-document for JSON.

---

[6] https://github.com/carml/carml.

[7] https://github.com/RMLio/RMLStreamer.

Fig. 4.13 Sensor
observations as table

ObservationID	Sensed Frequency	Sensor	Timestamp
64a93e56abb0	460	S1	1588684149
842483d5c22f	630	S2	1588684150
g46842483ds5	525	S2	1588684399
64a93e56abb0	460	S1	1588684149
842483d5c22f	630	S2	1588684150
64a93e56abb0	460	S1	1588684149
842483d5c22f	630	S2	1588684150
g46842483ds5	525	S2	1588684399
64a93e56abb0	460	S1	1588684149
842483d5c22f	630	S2	1588684150
g46842483ds5	525	S2	1588684399

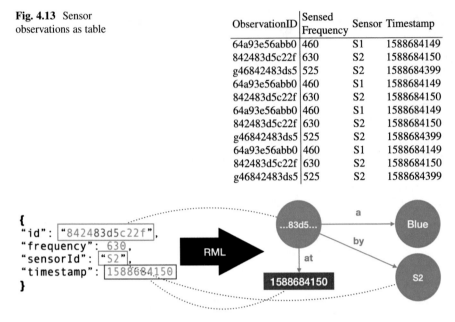

Fig. 4.14 The annotation process from a JSON stream to an RDF stream via RML

```
prefixes:
 color: ".../ontologies/colors#"
 fun: "http://.../idlab/function/"
mappings:
 color:
  sources:
   - ['colors.csv~csv']
  s: color:$(observation)
  po:
   - p: a
     o: color:Blue
     condition:
      function: fun:inRange
      parameters:
       - [fun:p_test,$(frequency)]
       - [fun:p_from,605]
       - [fun:p_to,650]
```

Listing 4.6 An RML Mapping.

4.3.5 Describe

This step of the life cycle aims at providing useful representations of the data
streams to be consumed by humans and/or Web agents. In practice, metadata
should be written using a formal language and the used terminology should be

shared. In the case of a dataset, the use of standard vocabularies for metadata management like DCAT or VOID is established. Metadata are then indexed by public data catalogs like DataHub, Zenodo, or Google's dataset search. The Web Data Commons indexes more than two million datasets annotated using schema.org. Nonetheless, the originality of streaming data calls for specific metadata. Open catalogs like dataset search struggle to interpret data semantics correctly, often falling in pitfalls that hinder the user's ability to discover data streams. To avoid such issues, two important questions should be answered:

? Questions to Answer

- How can we capture streaming data semantics?
- What metadata are essential for streaming data?

This section makes a parallelism with streaming linked data. In particular, it presents some prominent vocabularies that emerged to address the problem.

The **Vocabulary for Cataloging Linked Streams (VoCaLS)** (Prefix *voc:*) is an ontology to foster the interoperability between data streams and streaming services on the Web [34]. It consists of three modules for (1) publishing of streaming data following the Linked Data principles, (2) description of the streaming services that process the streams, and 3) tracking the provenance of stream processing [34].

The **Stream Annotation Ontology (SAO)** (Prefix *sao:*) allows publishing derived data about IoT streams. It is designed to represent both raw and aggregated data. The vocabulary allows to describe the aggregation transformations in-depth. SAO relies on PROV-O to track the aggregation provenance and OWL-Time for the temporal annotations [18].

Use Specialized Vocabularies to Describe Web Streams

Stream-specific metadata allow to correctly interpret the streaming data semantics, improving discovery. Moreover, they simplify the automation that concerns streaming linked data access.

Linked Data Event Stream (LDES)[8] (Prefix *ldes:*) defines a collection of immutable objects that evolve over time, describing both historical and real-time updates. *LDES* uses the TREE specification[9] for the modeling of the collections and data fragmentation purposes when the size of the collections becomes too big for a single HTTP response. TREE defines a collection of objects that adhere to a certain

---

[8] https://w3id.org/ldes/specification.

[9] https://w3id.org/tree/specification.

SHACL shape and how these collections can be fragmented and interlinked using multidimensional HTTP pagination [36].

Last but not least, **Schema.org (SORG)** already includes two concepts that are relevant for stream representation, i.e., DataFeed[10] and DataFeedItem.[11]

According to the FAIR initiative, metadata should be generous and extensive, including *descriptive* and *contextual* information about the data, as well as indications on data quality. However, the initiative gives some indication on what metadata are more relevant. The remainder of the section discusses the most important ones with respect to streaming linked data. In particular, the section shows how the aforementioned ontologies allow to (i) include descriptive information in natural language, e.g., title, published; (ii) include information about streaming data quality, e.g., rate and schema violations; (iii) present contextual domain knowledge, e.g., related ontologies; (iv) semantically annotate the publishing service; (v) present proper licensing (vi) include provenance metadata, e.g., generating pipeline or R2RML mappings.

Table 4.3 summarizes the analysis. Notably, the table summarizes also the explicit support for RDF Streams (I1) and Stream Descriptor (F3, A2), which are part of this book's recommendation. Moreover, the tables do not include evidence for I2 and I3. Indeed, none of the presented vocabularies fulfill I2, as they all rely on at least one non-FAIR import. On the other hand, all the ontologies support I3, since they all reuse concepts from other vocabularies.

Metadata (F2) *VoCALS* does not allow to include any information on data quality but limits its support to descriptive information about the resources, e.g., name and owner, and contextual information, e.g., the vocabulary used to annotate the stream content. *SAO* supports all the three metadata annotations. It is worth noticing the presence of specific classes and properties for annotating data quality by extending *QOI*.[12] *LDES* explicitly supports only contextual metadata as it directly relies on the TREE specification.

Service (A1) The FAIR prescription for serving data and metadata relies on standard protocols. While on the Web this usually means HTTP, it does not directly apply to streaming data that call for specific protocols. Except *LDES*, which inherits the HTTP access assumption from *TREE*, the other ontologies include a specific abstraction that aims at generalizing the access to the streaming data, i.e., `voc:StreamEndpoint`, `iots:Service`, `ces:EventService`.

---

[10] https://schema.org/DataFeed.

[11] https://schema.org/DataFeedItem.

[12] https://mobcom.ecs.hs-osnabrueck.de/cp_quality/.

Table 4.3 Summary of the 30,000 foot view analysis

Ontology	D	Q	C	RDF stream (I1)	Service (A1)	Descriptor (F3,A2)	LICENSE (R1)	Prov. (R2)
VoCALS	✓		✓	✓	✓	✓	Apache 2	✓
LDES			✓	≃		✓	CC	
SAO	✓	✓	✓		✓		CC	✓
SORG	✓	✓	✓		✓		CC	✓

✓ = supported, ≃ = partially supported, [D]escriptive, Data [Q]uality, [C]ontextual, [C]reative
[C]ommons

License (R1) All the selected ontologies have an explicit license. *VoCaLS* and *LDES* explicitly suggest associating a license with the annotated data streams.

Provenance (R2) Finally, tracking the provenance of the shared data is an encouraged practice from the FAIR initiative. In these regards, all the ontologies include dedicated classes and properties that allow to represent the analysis performed on the streaming data, i.e., voc:Query based on RSP-QL, CES ces:EventPattern for complex event recognition, sao:StreamAnalysis and iots:Analytics for continuous analysis of the data streams.

Example 4.4 (Cont'd) Carrying on the running example for description step, Listing 4.7 describes the color stream using one of the presented vocabularies, i.e., VoCaLS and DCAT The listing presents a stream descriptor, which is also a VoCaLS abstraction (see line 8). Descriptive metadata are listed between lines 10 and 12. On the other hand, contextual metadata about the stream, i.e., its landing page, are provided at line 15; service metadata are listed at line 16; and provenance is limited to the publisher (see line 14) and could be extended, for example, by describing the annotation process. Finally, service metadata are listed at line 16. More details are given in Listing 4.8, which is discussed in the next section.

> **Exercise**

Can you provide a similar descriptor for the original sensor stream? Can you extend the stream description included metadata about the annotation process, e.g., mapping or stream data vocabulary?

```
 1  PREFIX : <%\prefix%/resource/>
 2  PREFIX xsd: <http://www.w3.org/2001/XMLSchema#>
 3  PREFIX rdfs: <http://www.w3.org/2000/01/rdf-schema#>
 4  PREFIX dcat: <http://www.w3.org/ns/dcat#>
 5  PREFIX fmt: <http://www.w3.org/ns/formats/>
 6  PREFIX vocals: <http://w3id.org/rsp/vocals#>
 7  PREFIX vsd: <http://w3id.org/rsp/vocals-sd#>
 8  <> a vocals:StreamDescriptor .
 9
10  :colorstream a vocals:RDFStream ;
11    dcat:title "Color Stream"^^xsd:string ;
12    dcat:description "Stream of primary colors"^^xsd:string ;
13    dcat:license <https://creativecommons.org/licenses/by-nc/4.0/> ;
14    dcat:publisher "https://linkeddata.stream" ;
15    dcat:landingPage <http://linkeddata.stream/page/colorstream> ;
16    vocals:hasEndpoint :ColorEndpoint  .
```

Listing 4.7 Publishing color stream with Vocals and RSP-QL.

4.3.6 Serve

This step of the life cycle aims at making the streaming data accessible on the Web. The goal is to provision the data to the audience of interest, i.e., making them available for consumption. To this extent, three important questions should be answered:

? Questions to Answer

- What data should we provision?
- How should we provision such data?
- Who (or what) is responsible for the provisioning?

Looking back at the beginning of the life cycle, and in particular at the identification step, it emerges that two kinds of resources are relevant when publishing streaming data, i.e., *the stream itself and the resource it contains.* This idea, which was originally introduced by Barbieri et al. [5], suggests to decouple the sharing of streaming data and metadata. The former should be accessible in a continuous and reactive way, to promote streaming data access and streaming application development. The latter, instead, should be accessible via HTTP for backward compatibility with the Web infrastructure. Although the metadata document should be standalone, it should be also easy to link it to the actual streaming data, which are provisioned using a different yet more adequate protocol. To this extent, the stream descriptor should incorporate service metadata that describe the streaming data access.

Use HTTP for Metadata and Continuous Protocol for Streams

When publishing Web streams, it is essential to guarantee the reactivity of the stream content consumption as well as the backward compatibility of metadata retrieval.

Regarding specialized protocols, Van de Vyvere et al. wrote a comprehensive comparison between push- and pull-based protocols for Web data [35]. We briefly summarize below the most common protocols used for processing Web Streams [23]:

- **HTTP long polling** reduces the amount of requests by instructing the server to return responses to clients upon the occurrence of an update. However, it requires stateful resource management to maintain the connections open. HTTP long polling shows WebSockets-like performance when the underlying network latency is lower than half the data measurement rate [26].
- **Server-sent events** (SSE) extends, like HTTP long polling, and maintains an open connection open for every client but allows pushing multiple updates instead of one. SSE works best on HTTP/2, which multiplexes all requests and responses over one connection
- Last but not least, **WebSocket** is the most common protocol for real-time communication. After an HTTP handshake between client and server, it provides a bidirectional communication channel over one TCP connection for every client. WebSocket has a similar performance as SSE but a lower transmission latency when the server needs to send large messages, i.e., above 7.5 kilobytes. Also, for client to server communication, WebSocket provides a lower transmission latency than using HTTP [29].

In addition, several streaming oriented protocols exist (MQTT, CoAP [24], or STOMP[13]) with many advanced features, e.g., pub/sub via topic management. However, they are not extensively used in the context of Web Stream publishing. Provisioning data does not just concern the choice of a *protocol* but also the *data format*. The recommendation is to follow the standard guidelines to share the stream metadata (e.g., any suitable RDF serialization). On the other hand, for streaming data, the choice of the serialization might hugely impact the performance [12]. Moreover, Fernandez et al. [12] clarified the need for a form of punctuation for semantic stream that is called molecule. Due to the strict latency constraints of stream processing application, it is of paramount importance that the RDF stream serialization is in agreement with the molecule structure.

---

[13] https://stomp.github.io, Graph-QL Subscriptions.

Use RDF Serialization That Simplifies Time Annotation

In absence of a standard RDF serialization, adopt those ones that simplify the temporal annotation, i.e., JSON-LD and N-Quads.

In the absence of a standardized RDF stream serialization, existing works in the stream reasoning context seem to agree on adopting RDF formats that simplify the temporal annotation. In particular, the most common choice results in N-Quads and JSON-LD. Intuitively, both formats allow graph-based shaping to punctuate the stream. JSON-LD better fits semantically richer streams whose unit of information goes beyond the single triple. Nonetheless, the more expressive syntax allows to provision a context (which can also be referenced to avoid repeating information). N-Quads facilitates the parsing but requires to rebuild the graph unit in-memory, ultimately impacting the latency of complex processing.

Example 4.5 (Cont'd) Carrying on the running example for serving step, Listing 4.8 completes the stream descriptor presented in Listing 4.7 with the specification of the corresponding publishing service and access point. At line 4, the endpoint specifies the protocol, while at line 2, it specifies the format. In practice, multiple endpoints that provision alternative serialization of the stream are possible, following traditional content negotiation. Moreover, different protocols can be enabled to satisfy the needs of different clients. Listings 4.9 and 4.10 show the difference between the N-Quads and JSON-LD data formats, respectively.

```
:ColorEndpoint a vocals:StreamEndpoint ;
               dcat:format frmt:JSONLD ;
               dcat:accessURL "ws://colorstream:8080" ;
               dcat:protocol "WebSocket" .
```

Listing 4.8 An access point to the color stream.

```
<../64a93e56abb0> <../rdf-syntax-ns#type> <../colors#Red> <../colorstream
      #1588684149> .
<../842483d5c22f> <../rdf-syntax-ns#type> <../colors#Blue> <../colorstream
      #1588684150> .
<../g46842483ds5> <../rdf-syntax-ns#type> <../colors#Green> <../colorstream
      #1588684152> .
```

Listing 4.9 The RDF stream of color in N-Quads.

```
1   {   "@id":"http://linkeddata.stream/streams/colorstream#1588684149",
2       "@context":{
3           "color":"http://linkeddata.stream/ontologies/colors#",
4           "rdf":"http://www.w3.org/1999/02/22-rdf-syntax-ns#",
5           "rdfs":"http://www.w3.org/2000/01/rdf-schema#" },
6       "@graph":{   "@id":"64a93e56abb0", "@type":"color:Red" } },
7   {   "@id":"http://linkeddata.stream/streams/colorstream#1588684150",
8       "@context":{...},
9       "@graph":{   "@id":"842483d5c22f", "@type":"color:Blue" } },
10  {   "@id":"http://linkeddata.stream/streams/colorstream#1588684152",
11      "@context":{...},
12      "@graph":{   "@id":"f0e58ce1c485",   "@type":"color:Green" } }
```

Listing 4.10 The RDF stream of color in JSON-LD.

Finally, streaming data are exchanged across actors that expose, provision, and manipulate them. In traditional Web architecture, the notion of Agent or Service was dedicated to identify the actors interested or responsible for given resources. In the content of streaming linked data, a Web Stream Processing Service or Agent (if employing some sort of AI technique [31]) is a special kind of Web service that manipulates Web streams and, in particular, RDF Streams.

Definition 4.3 A Web Stream Processing (WSP) Agent/Service is a special kind of Web service that manipulates Web streams and, in particular, RDF Streams.

Three main types of Web stream services are relevant for the streaming linked data life cycle: (i) *Catalogs* that provide metadata about streams, their content, query endpoints, and more. (ii) *Publishers* that publish RDF streams, possibly following a Linked Data-compliant scheme (e.g., TripleWave in Listing 4.11). (iii) *Processors* that model a Stream Processing service, which performs any kind of transformation on streaming data, e.g., querying.

Describe Known Services That Are Related to Stream

Adding metadata about the services that are related to the stream helps contextualize the origin of the provision information as well as tracking the provenance of the data.

Intuitively, the first two services are relevant within streaming data publication, while the latter is relevant for processing and, thus, will be discussed in details in the next section. **Catalogs** are a common abstraction in the Web of Data due to the widespread need for dataset search. Nonetheless, existing data catalogs are not capable to capture the data stream semantics. Specialized RSP Catalogs can interpreter the semantic annotations (see Annotation Step) and enable streaming data discovery. The adoption of vocabularies like those mentioned above is a step toward solving this issue. **Publishers** deployed to provision the streaming data using a protocol suitable for continuous and/or reactive data access. As servers,

Publishers should implement content negotiation. Thus, publishers interested to serving RDF streams take over the conversion process. As discussed alongside the section, stream descriptors have a critical role in making the stream findable. To this extent, TripleWave [22] is a reusable and generic tool that enables the publication of RDF streams on the Web. It can be invoked through both pull- and push-based mechanisms, thus enabling RSP engines to automatically register and receive data from TripleWave.

Example 4.6 (Cont'd) Carrying on the running example, Listing 4.11 adds additional metadata for the stream publisher originally included in Listing 4.7, representing a bridge between publishers and catalog. Figure 4.15 shows how to carry on the life cycle consuming the sensor stream, converting it, and at the same time making the stream descriptor available.

```
<http://linkeddata.stream> a vsd:PublishingService ;
                    vsd:hasFeature vsd:transforming .
```

Listing 4.11 The publisher of the color stream.

4.3.7 Discovery and Access

After digging into the details of the processing step, it is worth noticing the gap between serving a processing in Fig. 4.3, i.e., between step (5) *Serve* and step (6) *Process*.

Figure 4.16 zooms in the life cycle to provide a better perspective on the gray area relating to the last two steps. Indeed, data stream *Discovery* and *Access*

Fig. 4.15 Converting and publishing the sensor stream with TripleWave

Fig. 4.16 Discovery, access, and analysis

are two intermediate steps that separate publication and processing. Despite their importance within the streaming linked data life cycle, most of the research that has been conducted in these two areas focuses on the publisher perspective (see Describe and Serve steps of the life cycle). In practice, data stream discovery and access are in their infancy and still require extensive ad hoc work for each use case. For these reasons, they are out of scope for this book, which aims at presenting only consolidated research work. On the other hand, the step that follows data processing is Analysis, which is often powered by sophisticated data visualization. The literature on this topic is extensive and best practices have been discussed in other volumes. Thus, they too are out of scope.

4.3.8 Process

The last step of the life cycle details how the data, which has been *Served* using the previous steps in the life cycle, can be processed. Looking at the literature on applied stream reasoning, one can notice that the processing steps tend to repeat. Although the community has not yet identified best practices, it is possible to enumerate a number of frequently used streaming data transformations:

- *Filter*, i.e., remove from the input streams the sub-portions that are not relevant for the analysis;
- *Enrich*, i.e., add to the input streams contextual data that make the analysis more informative;
- *Lift*, i.e., raise the input streams' abstraction level by means of background knowledge;
- *Merge*, i.e., combine two or more input streams to perform joint analyses;
- *Synthesize*, i.e., reduce the input streams to a summary by means of aggregations such as count, min, max, average, quartiles, or analytical function, e.g., Pearson correlation.

The remainder of the section first presents various stream reasoning frameworks and applications and then analyzes how they implement the processing practices listed above. Table 4.4 summarizes the analysis. Finally, to aid the comprehension, the section provides a visual representation using the color example.

BOTTARI [4] is a streaming analytic application designed to make sense of social media using deductive and inductive stream reasoning methods. The platform performs analyses of the activities of monitored influencers around the points of interest (POIs) of a given area. The analyses are window-based, spanning from few seconds to months. The social media streams are gathered from the Web (in particular from Twitter) and converted into an RDF stream using the proprietary crawling and sentiment mining infrastructure of Saltlux. BOTTARI, which employs augmented reality application for personalized and localized restaurant recommendations, was experimentally deployed in the Insadong district of Seoul.

Table 4.4 Summary of platforms processing features

Application	Domain	Filter	Enrich	Lift	Merge	Synthesize
BOTTARI [4]	Social media	σ, TP	POIs, Areas		Splice	Aggregates
SLD [3]	Event management	σ, TP	Sentiment		Splice	Aggregates
STAR-CITY [19]	Smart city	SPARQL	Map, Weather, Traffic history	OWL 2 EL	Splice	Aggregates, Prediction
CityPulse [27]	Smart city	σ, TP	Sensors	ASP	Cross	Aggregates, Traffic events
Agri-IoT [14]	Smart farm	σ, TP	Sensors	RDFS	Splice	Aggregates
Optique [15, 16]	Energy management	σ, π	Sensors configurations	DL-Lite$_A$	Cross, Splice	Aggregates

[T]riple [P]attern

The **CityPulse** project [27] handles a typical Smart City use case. The framework allows the development of applications that can provide a continuous and dynamic view of a city, making sense of social and sensor streams. To this extent, CityPulse employs semantic discovery, data analytics, and large-scale reasoning in real time. CityPulse is built in a service-oriented manner combining RDF Stream Processing and Complex Event Processing. The framework was demonstrated using live data from the city of Aarhus, Denmark.

SLD (which stands for Streaming Linked Data) [3] is a framework to collect, annotate, and analyze data streams. To this aim, SLD uses semantic technologies like RDF and SPARQL as well as techniques for sentiment mining. The framework follows the publication method proposed in [5]. SLD was successfully used to monitor the London Olympic Games 2012 and the Milano Design Week 2013 and 2016 editions.

STAR-CITY [19], which stands for Semantic Traffic Analytics and Reasoning for CITY, is a system for streaming data integration and analysis focusing on traffic data management. STAR-CITY is capable of interpreting the semantics of contextual information and then deriving innovative and easy-to-explore insights. In practice, it computes spatiotemporal similarities of traffic congestion and calculates accurate traffic forecasting using recent theoretical research work in contextual predictive reasoning. STAR-CITY was developed in collaboration with IBM in Dublin, Ireland, where it was also applied.

Agri-IoT [14] is a highly customizable online platform for IoT-based data analytics in the context of smart farming. It is capable of large-scale data processing and automatic reasoning based on data streams coming from a variety of sources, such as sensory systems, surveillance cameras, hyper-spectral images from drones, weather forecasting services, and social media. Agri-IoT aims at helping farmers make informed decision-making enabling prompt reactions to unpredictable events.

time

Fig. 4.17 Filtering blue colors

The **Optique** [15, 16] project handles several Big Data scenarios in the context of energy production. The case of Siemens Energy (Munich, Germany) aims at monitoring a number of service centers for power plants, whose main task is remotely monitoring thousands of appliances in real time, like gas and steam turbines, generators, and compressors installed within the plants. The case of Statoil (Stavanger, Norway) aims at improving the data gathering and analysis routines of Statoil geologists, who need IT specialists in order to make sense of multiple complex and large data sources. The Optique platform employs semantic technologies for enabling ontology-based data access for the aforementioned scenarios. Logical reasoning is used in form of query rewriting so to efficiently analyze data from heterogeneous data sources.

Streaming MASSIF [7] is an extension of the MASSIF platform for cascading stream reasoning. It combines several layers of processing that include RSP, Description Logic reasoning, complex event recognition, and aggregations. Streaming MASSIF was studied to overcome the limitations of MASSIF in terms of throughput and latency. Its layered structure eases the deployment of expressive components for advanced analytics.

Filter as Early as Possible

Filtering should precede other transformations to reduce the amount of data to be transferred.

Data streams are collected very close to their source at high frequency, with limited focus on data quality. Therefore, they are often filled with redundant, noisy, and nonrelevant data. Therefore, data cleansing is a necessary step for most of applications. In the mentioned frameworks, filtering is a stateless operation, i.e., it operates on a single stream element at a time. In these cases, filtering is possible using either traditional relational selection (σ) or using Triple pattern matching. More sophisticated forms of filtering are possible, e.g., STAR-CITY, which employs the full expressiveness of SPARQL to select data from a view over the input sources. An example of filtering from CityPulse requires to remove observations regarding parking with no vacancy or selecting locations with low congestion levels.

Example 4.7 Figure 4.17 presents the filtering practice for the color examples. From the stream of sensed colors, the filter removes non-blue observations and outputs a blue stream. Listing 4.12 shows the corresponding RSP-QL query.

```
1  PREFIX color: <http://linkeddata.stream/ontologies/colors#>
2  PREFIX :       <http://linkeddata.stream/resource/>
3  CONSTRUCT { ?b a color:Blue .}
4  FROM NAMED WINDOW <w> ON :colorstream [RANGE PT15S STEP PT5S]
5  WHERE {
6    WINDOW <w> { ?b a color:Blue .}
7  }
```

Listing 4.12   An example of Stream Filtering that selects all Blue occurrences.

Due to strict latency requirements, when publishing data streams, it is often a good idea to reduce the information to the minimum. However, sometimes it is necessary to put items in context by enriching streaming data with static ones. Indeed, static and slowly evolving data should not be repeatedly streamed; instead, they should be made available with streams' metadata. In CityPulse, Agri-IoT, and Optique, the positions and rates of the sensors evolve at a much slower rate than sensor observations. Therefore, it is reasonable to keep such metadata separate from the sensor measurements and assume them as static. Sensor metadata are used to enrich streaming observations with geospatial context, measure precision, or units of measure. Moreover, Optique also includes configurations regarding the platform as static metadata. Finally, BOTTARI and STAR-City show more complex cases of enrichment, with sensor data joined with POIs, traffic, and weather history.

Keep Static Data Small

To avoid performance loss in case of enrichment, keep the static data small, and, whenever necessary, apply cleansing techniques to avoid errors.

Example 4.8 One way to enrich the color stream is linking color and sentiment by leveraging the sentiment ontology (see Fig. 4.7). Figure 4.18 visualizes the example. The RSP-QL query to perform this enrichment is shown in Listing 4.13. The enrichment is done through the variable ?c that occurs both in the color stream and in the static sentiment graph.

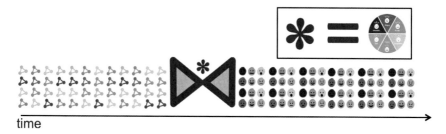

time

Fig. 4.18 Enriching the stream of color with sentiments

```
1  PREFIX s: <http://linkeddata.stream/ontologies/sentiment#>
2  CONSTRUCT{?color a  ?c; sentiment:hasSentiment ?sentiment. }
3  FROM NAMED <sentiment.rdf>
4  FROM NAMED WINDOW <rw>  ON :colorstream [RANGE PT30S STEP PT5S]
5  WHERE {
6    GRAPH <sentiment.rdf> { ?c :relates ?sentiment . }
7    WINDOW <rw> {
8        ?color a [ ?c rdfs:subClassOf color:Color .]
9    }
10 }
```

Listing 4.13 Enriching the color stream with its sentiment.

Similar to enrichment, lifting allows to leverage an external context for enhancing the analysis. In particular, logical reasoning can abstract the data in the stream to higher-level concepts defined in the domain knowledge. This can be useful for integration purposes. Considering the aforementioned frameworks, neither BOTTARI nor SLD performs any lifting. On the other hand, STAR-CITY and Agri-IoT adopt standard knowledge representation languages, i.e., OWL 2 EL and RDFS, to define knowledge bases. CityPulse uses Answer Set Programming to perform *lifting*. Domain knowledge is captured in the form of rules that are leveraged to raise the abstraction level of streaming data and operate on more complex concepts. Also Streaming MASSIF employs Description Logics to abstract event types and combines them using complex event recognition rules. Last but not least, reasoning is employed in Optique to perform query rewriting rather than lifting, which results to be more efficient in a streaming scenario due to the permanent nature of continuous queries. Nonetheless, the final result is comparable.

Little Semantics Goes the Long Way

Query rewriting is possible under certain conditions that limit the expressivity of the knowledge representation language. However, it gives substantial advantages in the streaming context, due to the long-lasting nature of continuous queries.

Fig. 4.19 Lifting the stream of colors

Example 4.9 In the color example, reasoning can be used to abstract the color to either *Warm* or *Cold* colors as shown in Fig. 4.19. Listing 4.14 shows the corresponding RSP-QL query that exploits the color ontology, as depicted in Fig. 4.8, to infer Warm colors.

```
1  PREFIX color: <http://linkeddata.stream/ontologies/colors#> .
2  PREFIX :       <http://linkeddata.stream/resource/> .
3  SELECT ?warm
4  FROM NAMED WINDOW <cw> ON :colorstream [RANGE PT15S STEP PT5S]
5  WHERE {
6    WINDOW <cw> { ?warm a color:WarmColor. }
7  }
```

Listing 4.14 Lifting warm colors.

All the aforementioned examples present multi-stream scenarios. Indeed, merging streams together is a standard practice in most commercial streaming applications. Without going too much in-depth of existing streaming join algorithms, this section distinguishes two main approaches to merging, i.e., Splice and Cross. The former consists in point-wise union of two streams without any temporal evaluation. This approach is common when multiple sources are annotated and converted locally, while data are rerouted within a mixed factory. Multiple streams can be combined to extend the expressivity of the analysis. Merging includes both splice, i.e., the element-wise union of two RDF streams, and crossing, i.e., time-based windowed joins. For example, streams from different parking lots can be *spliced* into one and crossed with one or more streams that monitor the streets in a span of the last 15 min. BOTTARI, SLD, STAR-CITY, and Agri-IoT focus on splicing, as they first merge all the input streams into a unified RDF stream. On the other hand, CityPulse and Optique allow to perform finer-grained analysis crossing only the most recent portions of the stream of interest.

Opt for Crossing When Streams Are Not Synchronized

When the streams have different rates, window-based merging is preferred as it regularizes the output.

Fig. 4.20 Color stream crossing using a window

Fig. 4.21 Color stream splicing

Example 4.10 Figures 4.20 and 4.21 show the difference between the two methods in terms of results. Listing 4.15 shows the corresponding RSP-QL query.

```
1 PREFIX color: <http://linkeddata.stream/ontologies/colors#>
2 PREFIX :       <http://linkeddata.stream/resource/> .
3 CONSTRUCT { ?g a color:Green ; color:from ?b, ?y . }
4 FROM NAMED WINDOW <bw> ON :bluestream [RANGE PT30S STEP PT5S]
5 FROM NAMED WINDOW <yw> ON :yellowstream [RANGE PT15S STEP PT5S]
6 WHERE {
7   WINDOW <bw> { ?b a color:Blue .}
8   WINDOW <yw> { ?y a color:Yellow .}
9   BIND( UUID() as ?g )
10 }
```

Listing 4.15 Creating a green stream by crossing blue and yellow ones.

Finally, synthesizing the activity on the streams is a first step toward streaming data analytics. In order to synthesize the content of a stream, RSP allows the use of *Aggregations*, which can consist of counting a number of occurrences, computing averages, computing minimum/maximum values, etc. Common aggregations are time-scoped for efficiency reasons, e.g., the average number of cars passing by a certain section within 15 min. Moreover, CityPulse suggests the reporting of composite events. In CityPulse, aggregations are used to compute the average vehicle speed over the past 5 min so to enable route planning. *Aggregations* are an important processing step in RSP, which allows to summarize the data that are captured inside a certain time window.

Use Time-Based Windows to Scope Advanced Practices

When it comes to the task of Synthesizing, using window-based computation can reduce the computational effort. In practice, window-based processing trades latency for throughput reducing the performance stress on the system.

time

Fig. 4.22 Color stream aggregation

Example 4.11 Figure 4.22 shows an example of an aggregation for the color example, i.e., counts the number of observations by color in the stream. Listing 4.16 extends the *Filtering* example and counts the number of blue colors in the window.

```
1   PREFIX color: <http://linkeddata.stream/ontologies/colors#>
2   PREFIX :      <http://linkeddata.stream/resource/>
3   SELECT (COUNT(?b) AS ?num_blue)
4   FROM NAMED WINDOW <w> ON :colorstream [RANGE PT15S STEP PT5S]
5   WHERE {
6   WINDOW <w> { ?b a color:Blue .}
7   }
```

Listing 4.16 An example of stream aggregation that counts all blue occurrences.

4.4   Chapter Summary

This chapter summarizes the steps of a streaming linked data life cycle. It discusses the following important points, i.e., identification of the streaming resources; data modeling with a focus on instantaneous, continuous, and time-varying concepts; schema management, a.k.a. shaping wrt. streaming data; annotation using mapping technologies for semantic extract-transform-load; and metadata management using vocabularies like DCAT and VoCALS. Finally, this chapter mentions the challenges in streaming data discovery and access, which lack the appropriate tools for managing data velocity. This chapter concludes with an overview of essential Web stream processing tasks, illustrated using the ColorWave example and implemented using RSP-QL.

References

1. Arasu, Arvind, Shivnath Babu, and Jennifer Widom. 2006. The CQL continuous query language: Semantic foundations and query execution. *VLDB Journal* 15 (2).
2. Balduini, Marco, and Emanuele Della Valle. 2015. Frappe: A vocabulary to represent heterogeneous spatio-temporal data to support visual analytics. In *International Semantic Web Conference (2)*. Vol. 9367. *Lecture Notes in Computer Science*, 321–328. Berlin: Springer.

3. Balduini, Marco, Emanuele Della Valle, Daniele Dell'Aglio, Mikalai Tsytsarau, Themis Palpanas, and Cristian Confalonieri. 2013. Social listening of city scale events using the streaming linked data framework. In *Proceedings of The Semantic Web – ISWC 2013 – 12th International Semantic Web Conference, Sydney, October 21–25, 2013, Part II,* ed. Harith Alani, Lalana Kagal, Achille Fokoue, Paul Groth, Chris Biemann, Josiane Xavier Parreira, Lora Aroyo, Natasha F. Noy, Chris Welty, and Krzysztof Janowicz, Vol. 8219. *Lecture Notes in Computer Science,* 1–16. Berlin: Springer.

4. Balduini, Marco, Irene Celino, Daniele Dell'Aglio, Emanuele Della Valle, Yi Huang, Tony Lee, Seon-Ho Kim, and Volker Tresp. 2012. Bottari: An augmented reality mobile application to deliver personalized and location-based recommendations by continuous analysis of social media streams. *Journal of Web Semantics* 16: 33–41.

5. Barbieri, Davide Francesco, and Emanuele Della Valle. 2010. A proposal for publishing data streams as linked data – A position paper. In *LDOW,* Vol. 628. *CEUR Workshop Proceedings.* CEUR-WS.org.

6. Bizer, Christian, Tom Heath, and Tim Berners-Lee. 2009. Linked data – the story so far. *International Journal on Semantic Web and Information Systems* 5 (3): 1–22.

7. Bonte, Pieter, Riccardo Tommasini, Emanuele Della Valle, Filip De Turck, and Femke Ongenae. 2018. Streaming MASSIF: Cascading reasoning for efficient processing of IoT data streams. *Sensors* 18 (11): 3832.

8. Calbimonte, Jean-Paul, Óscar Corcho, and Alasdair J. G. Gray. 2010. Enabling ontology-based access to streaming data sources. In *The Semantic Web – ISWC 2010 – 9th International Semantic Web Conference, ISWC 2010, Shanghai, November 7–11, 2010. Revised Selected Papers, Part I,* ed. Peter F. Patel-Schneider, Yue Pan, Pascal Hitzler, Peter Mika, Lei Zhang, Jeff Z. Pan, Ian Horrocks, and Birte Glimm. Vol. 6496. *Lecture Notes in Computer Science,* 96–111. Berlin: Springer.

9. Dell'Aglio, Daniele, Emanuele Della Valle, Jean-Paul Calbimonte, and Óscar Corcho. 2014. RSP-QL semantics: A unifying query model to explain heterogeneity of RDF stream processing systems. *International Journal on Semantic Web and Information Systems* 10 (4).

10. Dell'Aglio, Daniele, Emanuele Della Valle , Frank van Harmelen, and Abraham Bernstein. 2017. Stream reasoning: A survey and outlook. *Data Science* 1 (1–2): 59–83.

11. Ding, Li, Yun Peng, Paulo Pinheiro da Silva, Deborah L McGuinness, et al. 2005. Tracking RDF graph provenance using RDF molecules. TR-CS-05-06.

12. Fernández, Javier D., Alejandro Llaves, and Óscar Corcho. 2014. Efficient RDF interchange (ERI) format for RDF data streams. In *Proceedings of The Semantic Web – ISWC 2014 – 13th International Semantic Web Conference, Riva del Garda, October 19–23, 2014. Part II,* ed. Peter Mika, Tania Tudorache, Abraham Bernstein, Chris Welty, Craig A. Knoblock, Denny Vrandecic, Paul Groth, Natasha F. Noy, Krzysztof Janowicz, and Carole A. Goble. Vol. 8797. *Lecture Notes in Computer Science,* 244–259. Berlin: Springer.

13. Gao, Feng, Muhammad Intizar Ali, and Alessandra Mileo. 2014. Semantic discovery and integration of urban data streams. In *Proceedings of the Fifth Workshop on Semantics for Smarter Cities a Workshop at the 13th International Semantic Web Conference (ISWC 2014), Riva del Garda, October 19, 2014,* 15–30.

14. Kamilaris, Andreas, Feng Gao, Francesc X. Prenafeta-Boldu, and Muhammad Intizar Ali. 2016. Agri-IoT: A semantic framework for internet of things-enabled smart farming applications. In *3rd IEEE World Forum on Internet of Things, WF-IoT 2016, Reston, December 12–14, 2016,* 442–447. Washington: IEEE Computer Society.

15. Kharlamov, Evgeny, Dag Hovland, Martin G. Skjæveland, Dimitris Bilidas, Ernesto Jiménez-Ruiz, Guohui Xiao, Ahmet Soylu, Davide Lanti, Martin Rezk, Dmitriy Zheleznyakov, Martin Giese, Hallstein Lie, Yannis E. Ioannidis, Yannis Kotidis, Manolis Koubarakis, and Arild Waaler. 2017. Ontology based data access in statoil. *Journal of Web Semantics* 44: 3–36.

16. Kharlamov, Evgeny, Theofilos Mailis, Gulnar Mehdi, Christian Neuenstadt, Özgür L. Özçep, Mikhail Roshchin, Nina Solomakhina, Ahmet Soylu, Christoforos Svingos, Sebastian Brandt, Martin Giese, Yannis E. Ioannidis, Steffen Lamparter, Ralf Möller, Yannis Kotidis, and

Arild Waaler. 2017. Semantic access to streaming and static data at siemens. *Journal of Web Semantics* 44: 54–74.

17. Kharlamov, Evgeny, Yannis Kotidis, Theofilos Mailis, Christian Neuenstadt, Charalampos Nikolaou, Özgür Özcep, Christoforos Svingos, Dmitriy Zheleznyakov, Sebastian Brandt, Ian Horrocks, et al. 2016. Towards analytics aware ontology based access to static and streaming data. In *International Semantic Web Conference*, 344–362. Berlin: Springer.

18. Kolozali, Sefki, María Bermúdez-Edo, Daniel Puschmann, Frieder Ganz, and Payam M. Barnaghi. 2014. A knowledge-based approach for real-time IoT data stream annotation and processing. In *2014 IEEE International Conference on Internet of Things, Taipei, September 1–3, 2014*, 215–222.

19. Lécué, Freddy, Simone Tallevi-Diotallevi, Jer Hayes, Robert Tucker, Veli Bicer, Marco Luca Sbodio, and Pierpaolo Tommasi. 2014. Smart traffic analytics in the semantic web with STAR-CITY: scenarios, system and lessons learned in dublin city. *Journal of Web Semantics* 27–28: 26–33.

20. Lehmann, Jens, Spiros Athanasiou, Andreas Both, Alejandra García-Rojas, Giorgos Giannopoulos, Daniel Hladky, Jon Jay Le Grange, Axel-Cyrille Ngonga Ngomo, Mohamed Ahmed Sherif, Claus Stadler, Matthias Wauer, Patrick Westphal, and Vadim Zaslawski. 2015. Managing geospatial linked data in the geoknow project. In *The Semantic Web in Earth and Space Science. Current Status and Future Directions*, ed. Tom Narock and Peter Fox. Vol. 20. *Studies on the Semantic Web*, 51–78. Amsterdam: IOS Press.

21. Margara, Alessandro, Jacopo Urbani, Frank van Harmelen, and Henri E. Bal. 2014. Streaming the web: Reasoning over dynamic data. *Journal of Web Semantics* 25: 24–44.

22. Mauri, Andrea, Jean-Paul Calbimonte, Daniele Dell'Aglio, Marco Balduini, Marco Brambilla, Emanuele Della Valle, and Karl Aberer. 2016. Triplewave: Spreading RDF streams on the web. In *ISWC*.

23. Murley, Paul, Zane Ma, Joshua Mason, Michael Bailey, and Amin Kharraz. 2021. Websocket adoption and the landscape of the real-time web. In *WWW '21: The Web Conference 2021, Virtual Event/Ljubljana, Slovenia, April 19–23, 2021*, ed. Jure Leskovec, Marko Grobelnik, Marc Najork, Jie Tang, and Leila Zia, 1192–1203. New York: ACM/IW3C2.

24. Naik, Nitin. 2017. Choice of effective messaging protocols for IoT systems: MQTT, CoAP, AMQP and HTTP. In *2017 IEEE International Systems Engineering Symposium (ISSE)*, 1–7.

25. Ngomo, Axel-Cyrille Ngonga, Sören Auer, Jens Lehmann, and Amrapali Zaveri. 2014. Introduction to linked data and its lifecycle on the web. In *Proceedings of the Reasoning Web. Reasoning on the Web in the Big Data Era – 10th International Summer School 2014, Athens, September 8–13, 2014*, ed. Manolis Koubarakis, Giorgos B. Stamou, Giorgos Stoilos, Ian Horrocks, Phokion G. Kolaitis, Georg Lausen, and Gerhard Weikum. Vol. 8714. *Lecture Notes in Computer Science*, 1–99. Berlin: Springer.

26. Pimentel, Victoria, and Bradford G. Nickerson. 2012. Communicating and displaying real-time data with websocket. *IEEE Internet Computing* 16 (4): 45–53.

27. Puiu, Dan, Payam M. Barnaghi, Ralf Toenjes, Daniel Kuemper, Muhammad Intizar Ali, Alessandra Mileo, Josiane Xavier Parreira, Marten Fischer, Sefki Kolozali, Nazli FarajiDavar, Feng Gao, Thorben Iggena, Thu-Le Pham, Cosmin-Septimiu Nechifor, Daniel Puschmann, and João Fernandes. 2016. Citypulse: Large scale data analytics framework for smart cities. *IEEE Access* 4: 1086–1108.

28. Sequeda, Juan F., and Óscar Corcho. 2009. Linked stream data: A position paper. In *SSN*. Vol. 522. *CEUR Workshop Proceedings*, 148–157. CEUR-WS.org.

29. Slodziak, Wojciech, and Ziemowit Nowak. 2015. Performance analysis of web systems based on xmlhttprequest, server-sent events and websocket. In *Information Systems Architecture and Technology: Proceedings of 36th International Conference on Information Systems Architecture and Technology – ISAT 2015 – Part II, Karpacz, September 20–22, 2015*, ed. Adam Grzech, Leszek Borzemski, Jerzy Swiatek, and Zofia Wilimowska. Vol. 430. *Advances in Intelligent Systems and Computing*, 71–83. Berlin: Springer.

30. Stonebraker, Michael, Ugur Çetintemel, and Stanley B. Zdonik. 2005. The 8 requirements of real-time stream processing. *SIGMOD Record* 34 (4): 42–47.

31. Tommasini, Riccardo, Davide Calvaresi, and Jean-Paul Calbimonte. 2019. Stream reasoning agents: Blue sky ideas track. In *AAMAS*, 1664–1680. Stanbul: International Foundation for Autonomous Agents and Multiagent Systems.
32. Tommasini, Riccardo, Mohamed Ragab, Alessandro Falcetta, Emanuele Della Valle, and Sherif Sakr. 2020. A first step towards a streaming linked data life-cycle. In *Proceedings of The Semantic Web – ISWC 2020 – 19th International Semantic Web Conference, Athens, November 2–6, 2020, Part II*, ed. Jeff Z. Pan, Valentina A. M. Tamma, Claudia d'Amato, Krzysztof Janowicz, Bo Fu, Axel Polleres, Oshani Seneviratne, and Lalana Kagal. Vol. 12507. *Lecture Notes in Computer Science*, 634–650. Berlin: Springer.
33. Tommasini, Riccardo, Pieter Bonte, Femke Ongenae, and Emanuele Della Valle. 2021. RSP4J: an API for RDF stream processing. In *Proceedings of The Semantic Web – 18th International Conference, ESWC 2021, Virtual Event, June 6–10, 2021*, ed. Ruben Verborgh, Katja Hose, Heiko Paulheim, Pierre-Antoine Champin, Maria Maleshkova, Óscar Corcho, Petar Ristoski, and Mehwish Alam. Vol. 12731. *Lecture Notes in Computer Science*, 565–581. Berlin: Springer.
34. Tommasini, Riccardo, Yehia Abo Sedira, Daniele Dell'Aglio, Marco Balduini, Muhammad Intizar Ali, Danh Le Phuoc, Emanuele Della Valle, and Jean-Paul Calbimonte. 2018. Vocals: Vocabulary and catalog of linked streams. In *International Semantic Web Conference (2)*. Vol. 11137. *Lecture Notes in Computer Science*, 256–272. Berlin: Springer.
35. Van de Vyvere, Brecht, Pieter Colpaert, and Ruben Verborgh. 2020. Comparing a polling and push-based approach for live open data interfaces. In *Proceedings of theWeb Engineering – 20th International Conference, ICWE 2020, Helsinki, June 9–12, 2020*, ed. Mária Bieliková, Tommi Mikkonen, and Cesare Pautasso. Vol. 12128. *Lecture Notes in Computer Science*, 87–101. Berlin: Springer.
36. Van Lancker, Dwight, Pieter Colpaert, Harm Delva, Brecht Van de Vyvere, Julián Andrés Rojas Meléndez, Ruben Dedecker, Philippe Michiels, Raf Buyle, Annelies De Craene, and Ruben Verborgh. 2021. Publishing base registries as linked data event streams. In *Proceedings of the Web Engineering – 21st International Conference, ICWE 2021, Biarritz, May 18–21, 2021*, ed. Marco Brambilla, Richard Chbeir, Flavius Frasincar, and Ioana Manolescu. Vol. 12706. *Lecture Notes in Computer Science*, 28–36. Berlin: Springer.
37. Wilkinson, Mark D., Michel Dumontier, IJsbrand Jan Aalbersberg, Gabrielle Appleton, Myles Axton, Arie Baak, Niklas Blomberg, Jan-Willem Boiten, Luiz Bonino da Silva Santos, Philip E. Bourne, Jildau Bouwman, Anthony J. Brookes, Tim Clark, Mercè Crosas, Ingrid Dillo, Olivier Dumon, Scott Edmunds, Chris T. Evelo, Richard Finkers, Alejandra Gonzalez-Beltran, Alasdair J. G. Gray, Paul Groth, Carole Goble, Jeffrey S. Grethe, Jaap Heringa, Peter A. C 't Hoen, Rob Hooft, Tobias Kuhn, Ruben Kok, Joost Kok, Scott J. Lusher, Maryann E. Martone, Albert Mons, Abel L. Packer, Bengt Persson, Philippe Rocca-Serra, Marco Roos, Rene van Schaik, Susanna-Assunta Sansone, Erik Schultes, Thierry Sengstag, Ted Slater, George Strawn, Morris A. Swertz, Mark Thompson, Johan van der Lei, Erik van Mulligen, Jan Velterop, Andra Waagmeester, Peter Wittenburg, Katherine Wolstencroft, Jun Zhao, and Barend Mons. 2016. The fair guiding principles for scientific data management and stewardship. *Scientific Data* 3 (1): 160018.

Chapter 5
Web Stream Processing Systems and Benchmarks

Abstract RDF Stream Processing aims at taming data velocity and variety simultaneously. In addition to solid theoretical foundation, the research community has shown its appreciation for working prototypes that help investigating the Stream Reasoning research question empirically. Nonetheless, empirical research is hard: the manual effort to set up the experimental environment is vast and error-prone, the experiments are often not documented and hard to reproduce, and working prototypes may have different dependencies that increase the incompatibility. This chapter presents the community efforts to tame the challenges above and mitigate the problem of a fair performance evaluation for RDF Stream Processing engines. In particular, this chapter presents systems architectures, benchmarks, and tool for RSP benchmarking.

5.1 Introduction

RDF Stream Processing aims at taming data velocity and variety simultaneously. In particular, previous chapters have shown how these challenges impact a number of use cases. Therefore, the community behind RSP, in addition to solid theoretical foundations, values the role of working prototypes to enable empirical research as shown by Fig. 5.1.

Despite several attempts to follow software engineering best practices, researchers tend to design architectures in isolation and develop prototypes independently. Therefore, comparative evaluation presents many challenges: (i) the manual effort to set up the experimental environment is vast and error-prone; (ii) experiments are often not documented and hard to reproduce; (iii) prototypes may have different dependencies that increase the incompatibility.

The community efforts to reduce the impact of the challenges described above include benchmarks and evaluation methodologies. Nonetheless, reproducibility and repetitiveness of the evaluations are not always guaranteed. Indeed, when the complexity of the evaluation is high, hypotheses testing becomes even harder, e.g., results are situational, no absolute winner emerges, and it may not be clear under which conditions an approach outperforms another [31].

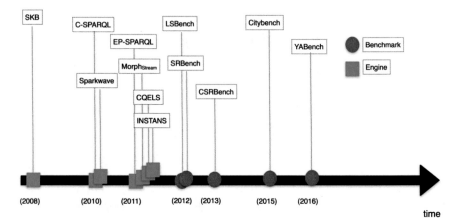

Fig. 5.1 Timeline of RDF stream processing engines and benchmarks (incomplete)

This chapter presents the community efforts to study the performance of RDF Stream Processing engines. In particular, this chapter takes into account the achievements of the RSP W3C community group,[1] which is working toward the standardization of a data model, i.e., *RDF Streams*, and of a query language, i.e., RSP-QL [14]. To this extent, this chapter surveys existing RSP Engines, focusing on the most prominent ones, i.e., C-SPARQL engine, CQELS, SPARQL$_{Stream}$, and surveys various benchmarks [1, 15, 19, 24, 36] and efforts to foster empirical research [31, 32].

5.2   RSP Engines (And Their Alignment with RSP-QL)

Before the introduction of RSP-QL, a variety of RSP languages emerged over time, e.g., C-SPARQL [6], CQELS-QL [21], SPARQL$_{stream}$ [10], and Strider-QL [25]. Such languages that support some form of continuous semantics are extensions of SPARQL. RSP languages are usually paired with working prototypes that helped proving the feasibility of the approach, as well as studying its efficiency. This section details the similarities and differences between existing RSP engines and how they align to RSP-QL.

---

[1] http://www.w3.org/community/rsp/.

5.2.1   Evaluation Time

When developing Stream Processing Engines to evaluate continuous queries, there are a number of design decisions that might impact the query correctness. Such decisions, which are usually hidden in the query engine implementation, define the so-called *operational semantics* (also known as *execution semantics*). With their SECRET model, Botan et al. [9] identified a set of four primitives that formalize the operational semantics of window-based stream processing engines. RSP-QL incorporates these primitives and applies them on existing RSP-QL engines:

- *Scope* is a function that maps an event-time instant t_e to the temporal interval where the computation occurs.
- *Content* is a function that maps a processing time instant t_p to the subset of stream elements included in the interval identified by the scope function.
- *Report* is a dimension that characterizes under which conditions the stream processors emit the window content. SECRET defines four reporting dimensions:

  - (**CC**) *Content Change*: the engine reports when the content of the current window changes.
  - (**WC**) *Window Close*: the engine reports when the current window closes.
  - (**NC**) *Non-empty Content*: the engine reports when the current window is not empty.
  - (**P**) *Periodic*: the engine reports periodically.

  Note that combinations of reporting dimensions are possible. For example, the C-SPARQL engine reports under *Content Change* and *Non-empty Content*
- *Tick* is a dimension that explains what triggers the report evaluations. Possible ticks are time-driven, tuple-driven, or batch-driven.

5.2.2   C-SPARQL

The C-SPARQL Engine [6] is an RSP engine that adopts a black box approach by pipelining a DSMS system with a SPARQL engine. The DSMS is used to execute the S2R operators and the execution semantics, while the SPARQL engine performs the evaluation of the queries implemented as the R2R operator. C-SPARQL supports the Window Close and Non-empty Content reporting policies while employing RStreams as R2S operators. The C-SPARQL language is an extension of SPARQL 1.0, with the addition of the continuous operators. An example of the C-SPARQL query to process the color stream can be found in Listing 5.1.

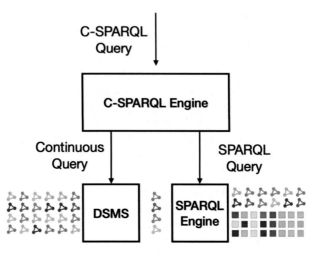

Fig. 5.2 C-SPARQL engine architecture

```
1  REGISTER QUERY GreenColors AS
2  PREFIX color: <http://linkeddata.stream/ontologies/colors#>
3  PREFIX :     <http://linkeddata.stream/resource/>
4  SELECT ?green
5  FROM STREAM :colorStream [RANGE 5s STEP 1s]
6  WHERE {
7    ?green a color:Green.
8  }
```

Listing 5.1 Example of the C-SPARQL query to process the color stream.

Figure 5.2 shows an abstract view of the engine architecture. C-SPARQL engine allows its users to register a continuous query written using C-SPARQL syntax. Then the engine splits the query in two subparts that are delegated to independent components. A Data Stream Management System (DSMS) is responsible for the continuous query evaluation. It deals with the input stream using window operators. A SPARQL engine operates the remainder part of the query, considering every window update. The query result is finally formatted according to the query clause.

5.2.3 CQELS

The CQELS Engine [21] takes a white box approach, such that it has access to all the available operators, allowing the optimization of the query evaluation. Compared to C-SPARQL, it supports the Content Change reporting strategy. Furthermore, CQELS supports the IStream R2S operator instead of the RStream. Furthermore, the S2R operator directly evaluates Triple Patterns and thus produces Time-Varying Bindings instead of Time-Varying Graphs. CQELS also supports various optimizations, such as adaptively reordering the query execution plan in order to optimize

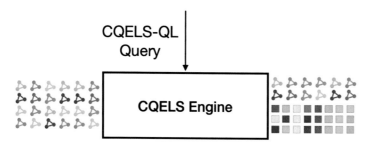

Fig. 5.3 CQELS engine architecture

throughput and encoding of the triples, to improve memory consumption. It is important to note that since CQELS supports the Content Change reporting strategy, the query is evaluated each time a triple arrives, while C-SPARQL processes the new triples in batches due to the Window Close policy. Similar to C-SPARQL, CQELS extends SPARQL 1.1 with temporal operators. An example of the CQELS query to process the color stream can be found in Listing 5.2.

```
1  PREFIX color: <http://linkeddata.stream/ontologies/colors#>
2  PREFIX :       <http://linkeddata.stream/resource/>
3  SELECT ?green
4  WHERE {
5  STREAM :colorStream [RANGE 5s]
6    {?green a color:Green.}
7  }
```

Listing 5.2 Example of the CQELS query to process the color stream.

Figure 5.3 shows the abstract architecture of the CQELS Engine that, differently from C-SPARQL engines, is a native RDF Stream Processing system. Indeed, the CQELS engine natively implements streaming operators and supports adaptive query optimization. During the query execution, the engine can reorder the operators to maximize the efficiency.

5.2.4 SPARQL$_{stream}$

SPARQL$_{stream}$ [10] focuses on querying virtual RDF streams with SPARQL$_{stream}$. Thus, compared to C-SPARQL and CQELS, it uses Ontology-Based Data Access to virtually map raw data to RDF data. Similarly to C-SPARQL, it supports the Window Close and Non-empty Content reporting policies. Morph$_{stream}$ is the only engine that supports all R2S operators. An example of the SPARQL$_{stream}$ query to process the color stream can be found in Listing 5.3.

Fig. 5.4 Morph*Stream*
architecture

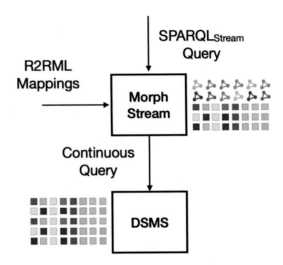

```
1  PREFIX color: <http://linkeddata.stream/ontologies/colors#>
2  PREFIX :      <http://linkeddata.stream/resource/>
3  SELECT ?green
4  FROM NAMED STREAM :colorStream [NOW-5 MINUTES TO NOW-0 MINUTES]
5  WHERE {
6    ?green a color:Green.
7  }
```

Listing 5.3 Example of the SPARQL stream query to process the color stream.

Figure 5.4 shows the Morph$_{stream}$ architectural schema, which highlights a secondary input (in addition to the query) that represents the mapping file. The rewriting process can target a variety of DSMSs, e.g., SNEE or Esper and Sensor Middleware. The resulting data streams are then mapped to the original engine to output a graph stream, in case the construct clause is specified.

5.2.5   Strider

Strider [25] is a hybrid adaptive distributed RSP engine that optimizes the logical query plan according to the state of the data streams. It is built upon Spark Streaming and borrows most of its operators directly from Spark. Strider translates Strider-QL queries to Spark Streaming's internal operators. It inherits the Window Close reporting policy from Spark Streaming and supports the RStream as R2S operator. Listing 5.4 shows an example query in Strider-QL. Note that the additional *BATCH* keyword is used for the definition of micro-batches for the underlying Spark Streaming.

```
 1 STREAMING { WINDOW [ 20 SECONDS ] SLIDE [ 20 SECONDS ] BATCH [ 5 SECONDS ] }
 2 REGISTER { QUERYID [ ColorQuery ]
 3   SPARQL [
 4   PREFIX color: <http://linkeddata.stream/ontologies/colors#>
 5   PREFIX :      <http://linkeddata.stream/resource/>
 6   SELECT ?green
 7   WHERE {
 8     ?green a color:Green.
 9     }
10   ]
11 }
```

Listing 5.4 Example of the Strider-QL query to process the color stream.

Figure 5.5 shows the engine architecture. Due to its distributed nature, Strider is more complex than the other engines presented above. The mentioned figure presents two parts: (1) above the blue dashed line one can observe the application's data flow management, which consists of a pub/sub that collects IoT streams. The receiving layer in turn emits messages for the streaming layer that executes the continuous query (2) the lower part of the figure shows the components of the computing core. The first component registers the continuous query and—after being parsed—gets statically optimized. At last, the Query Processing layer ensures that the query execution goes smoothly with reference to the optimized logical plan.

5.2.6 Comparison

In addition to the rigid yet explicit characteristics that each engine holds, it's also worth mentioning how engines show inherent subtle differences. For example, none of them allows to define the starting timestamp t^0 as part of the time-based sliding window operators definition. This means that the starting time is supplied by the engine itself and—in the case of processing time—engines are bound to produce different results. Differences like the t^0 make impossible to correctly compare the results produced by the various ones. Table 5.1 summarizes the differences between the engines. Note that RSP4J [33] supports the configuration of all these parameters, promoting comparable research.

5.2.7 Related Works

Additional RSP engines were proposed, e.g., Streaming knowledge bases [34], INSTANS [26], SparkWave [20] and IMaRS [12]. These engines were not included in the discussion above as their prototypes are no longer available for evaluation. Nonetheless, they are relevant for SR research.

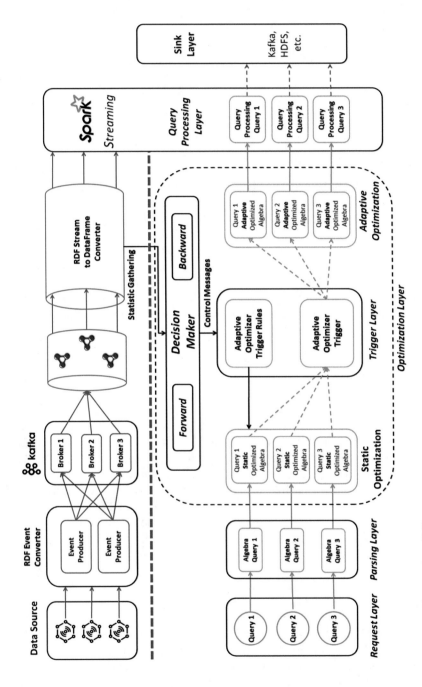

Fig. 5.5 Strider architecture from [25]

Table 5.1 Summary of the differences between RSP engines. $TVG_{W,S}$ denotes the time-varying graph and G_0 the default graph

Feature/engine	C-SPARQL	CQELS	$SPARQL_{Stream}$	Strider
RSP-QL dataset	$TVG_{W,S}$ in G_0	Named $TVG_{W,S}$	$TVG_{W,S}$ in G_0	$TVG_{W,S}$ in G_0
Report	WC, NC	CC	WC, NC	WC
R2S	RStream	IStream	RStream, IStream and DStream	RStream

INSTANS differs from other engines because it does not rely on window operations. Instead, a continuous query problem is modeled as multiple inter-connected SPARQL queries and rules. The engine performs incremental query evaluation with reference to the incoming RDF data against the compiled set of queries, storing intermediate results [27]. Finally *IMaRS* and *SparkWave* attempt to go beyond SPARQL expressiveness including incremental RDFS reasoning. The former focuses on the intuition that deletions can be predicted when maintenance is regulated by a window operator [12], while the latter extends the Rete Algorithm for incremental graph computation to support windowing [20].

5.2.8   CEP Enabled RSP Engines

EP-SPARQL [3] is a SPARQL 1.0 extension for complex event processing over RDF Streams (see Listing 5.5 for an example). It supports the temporal operators SEQ and EQUALS, with their OPTIONAL variants, and Allen's algebra relations [2].

EP-SPARQL queries are translated into Prolog rules and evaluated accordingly with the ETALIS execution model, which is based on event-driven backward chaining (EDBC). Efficient rule evaluation relies on (left-associative) rules binarization, i.e., binary intermediate goals. Indeed, binary rules are easier to compute and to manage and encourage goal sharing across queries.

EP-SPARQL consumes streams of timestamped RDF triples as input, encoded as Datalog predicates $triple(s,o,p,t_s,t_e)$, where t_s,t_e indicate the starting and ending timestamp of the triple, respectively. EP-SPARQL relies on time monotonicity, i.e., events are assumed to arrive in order. Moreover, it is downward compatible with SPARQL 1.0; this practically means that any SPARQL 1.0 query is also a valid EP-SPARQL query. Basic Graph Pattern (BGP) is translated into a Prolog predicate $triple(s, p, o)$, and results are computed using extended SPARQL 1.0 solution mappings.[2]

As a consequence of ETALIS' computational model, graph pattern-matching and event pattern-matching happen at attribute level, and EP-SPARQL can handle and seamlessly evaluate them. Moreover, EP-SPARQL can answer queries in

---

[2] Notably, mapping compatibilities rely on timestamps.

combination with background knowledge encoded as additional Prolog rules that are considered during the evaluation. Last but not least, ETALIS provides alternative execution policies under which EP-SPARQL queries may be evaluated, i.e., recent, naive, chronological, and unrestricted

Listing 5.5 presents an example query using a preliminary EP-SPARQL syntax. The query is checking for sequences of colors: first greens are followed by blues and then reds followed by yellows.

```
1 PREFIX color: <http://linkeddata.stream/ontologies/colors#>
2 SELECT ?green
3 WHERE {
4     {?green a color:Green. } SEQ {?blue a color:Blue. }
5     SEQ
6     {?red a color:Red. } SEQ {?yellow a color:Yellow. }
7     }
8 FILTER ( getDURATION() < "P1H"^^xsd:duration)
```

Listing 5.5 EP-SPARQL query for detecting sequences of colors.

RSEP-QL [16] is an extension of RSP-QL [13] for Complex Event Processing. Like RSP-QL, it aims at explaining the behavior of existing solutions, i.e., EP-SPARQL and C-SPARQL. Thus, RSEP-QL unifies event detection and analytics operations for RDF Stream Processing.

RSEP-QL introduces Window Functions to generalize the time-based window operators of RSP-QL. Window functions can be time-preserving and enable time-based comparisons and, thus, complex event processing. A window function W takes a vector parameter **p** and an input stream S to produce a finite substream S' of S denoted by W[**p**](S) [16].

Furthermore, RSEP-QL introduces Event Patterns (EP) for computing complex event processing. EPs include:

- Basic Event Patterns that are labeled and defined contextually to a named window w, i.e., (EVENT w P), where P is an RSP-QL Basic Graph Pattern.
- (Complex) Event Patterns, i.e., (FIRST E_i), (LAST E_j), and (E_i SEQ E_j) where E_i and E_j are event patterns.
- Event Graph Pattern, denoted as MATCH E_i where E_i is an event pattern.

Event Patterns evaluation semantics extends RSP-QL evaluation function. It is worth mentioning that the evaluation of an Event Graph Pattern removes the time annotation and returns a bag of solution mappings. Thus, the evaluation of an Event Graph Pattern isolates the Complex Event Processing features from the RSP-QL ones.

As indicated in [16], RSEP-QL captures EP-SPARQL behavior including selection and consumption policies as alternative semantics for the *SEQ* operators. Although RSEP-QL was not designed as a query language but as a reference model, Listing 5.6 presents an example query using a preliminary RSEP-QL syntax. The listing also looks for sequences of colors, greens followed by blues. Note that it is not looking for the additional sequence of reds and yellows, as this is very tricky to define in RSEP-QL.

```
1  PREFIX color: <http://linkeddata.stream/ontologies/colors#>
2  REGISTER <ColorSequences> AS
3  CONSTRUCT { ?green color:hasNextColor ?blue. }
4  FROM NAMED :S WIN [LND 1H] AS :w1
5  {?green a color:Green. } SEQ {?blue a color:Blue. }
6  EVENT ON :w1 { ?green a color:Green.} AS green
7  EVENT ON :w1 { ?blue a color:Blue. } AS blue
8  WHERE   { MATCH { green SEQ blue} }
```

Listing 5.6 RSEP-QL query for detecting sequences of colors. (Partial)

Ontology-based Event Processing (**OBEP**) [30] seamlessly combines temporal operators with a family of knowledge representation languages around the notion of event. It allows Web users to specify and compose events working with high-level abstractions over semantic streams while modeling events as First Class Citizens (FCC). As events are FCC, it allows to (1) name and compare events, (2) pass events as arguments to various temporal operators and expressions, and (3) return events as results of the evaluation of the operators.

OBEP allows to define logical events, through Description Logics definitions in Manchester Syntax,[3] which has shown to be very concise and readable. As the reasoning for the inference of these logical events is orthogonal to the detecting of temporal patterns on top of these logical events, expressive reasoning paradigms, such as OWL2 reasoning, can be used for defining and inferring these logical events.

OBEP introduces Composite Event Expressions to combine logical events through temporal operators, such as FIRST, LAST, AND, and SEQ. These are similar to the once supported in RSEP-QL. Furthermore, OBEP also supports Allen's interval relations. Allen's Algebra is a well-known formalism that specifies 12 binary relations that are possible w.r.t. a pair of intervals. These relations include BEFORE, MEETS, OVERLAPS, DURING, STARTS, FINISHES, and their inverse. To enable filters on Composite Event Expressions, OBEP supports various Data Guards. These guards allow to filter or join the events contained in the more abstract definitions.

Listing 5.7 shows the same color sequence example as for EP-SPARQL and RSEP-QL, where it detects sequences of greens followed by blues, which are followed by sequences of reds followed by yellows.

---

[3] https://www.w3.org/TR/owl2-manchester-syntax/.

```
 1  FROM STREAM <http://linkeddata.stream/resource/colorStream> WITH <http://
       linkeddata.stream/ontologies/colors#>.
 2
 3  EVENT GreenEvent AS some Green.
 4  EVENT BlueEvent AS some Blue.
 5  EVENT RedEvent AS some Red.
 6  EVENT YellowEvent AS some Yellow.
 7
 8  EVENT GreenBlue MATCH GreenEvent SEQ BlueEvent WITHIN 1H
 9  EVENT RedYellow MATCH RedEvent SEQ YellowEvent WITHIN 1H
10  EVENT ColorSequence MATCH (GreenBlue SEQ RedYellow) WITHIN 1H
11
12  RETURN ColorSequence AS RDF STREAM
```

Listing 5.7 OBEP query for detecting sequences of colors.

5.3   Benchmarking Web Stream Processing

The section introduces the challenges in evaluating Web Stream Processing systems and briefly describes the domain-specific benchmarks that were released over the years. At the time of writing, this area of research is very active. Therefore, the section focuses on most prominent and accepted benchmarks, briefly discussing their characteristics. Finally, the section presents the community attempts to make the WSP performance research systematic and comparative. Indeed, the WSP community has worked hard to design test drivers that facilitate the benchmark execution.

5.3.1   Benchmarking Challenges and Key Performance Indicators

The problem of benchmarking WSP systems, with the purpose of defining a generic methodology to evaluate their performance, was initially developed by Scharrenbach et al. [29]. The authors identified the challenges that WSP systems must face and discerned the key performance indicators (KPIs) to be measured in order to describe the performance of such WSP engines.

Referring to the generic Information Flow Processor architecture (see Chap. 2, Sect. 2.2.3), Scharrenbach et al. identify five challenges (C) and describe their impact in terms of WSP system properties, i.e., (C1) **Background data management** refers to the system ability to combine static and streaming data. The challenge, which also impacts the data distribution of the processing strategy, derives from the need for rapidly accessing static data without relying extensively on indexes. (C2) **Expressiveness**: the support of logical inference is one of the unique characteristics of WSP systems. However, the expressiveness/efficiency trade-off becomes even more challenging when data change rapidly. It requires the use of background knowledge, e.g., rules; moreover, it forces the adoption of semantically annotated data streams and a time model that keeps the inference complexity low. Indeed, (C3) the **time model** has a huge impact on the system performance. For example, interval-based semantics is harder to manage than point-based one [35]. (C4) **Processing**, and in particular continuous query answering, requires to choose an appropriate strategy for storing, accessing, and discarding partial results. In the presence of static data, an efficient caching and memory management is essential. Additionally, focusing just on those elements of the stream that are relevant for processing is a key challenge. An inappropriate choice, e.g., too wide windows, may negatively impact the system performance and the quality of service. Other related challenges rise from reporting of query results as well as the concurrent management of multiple queries. (C5) **Response to burst** is the system capability to maintain the quality of service in the presence of sudden increases of the input stream rate. Indeed, WSP systems are designed to operate under continuous semantics. However, the amount of the input stream rate is often unpredictable. For instance, the number of tweets per minute varies with the popularity of a topic. Managing bursts is hard, especially in a distributed context. Indeed, it requires a careful planning of the processing tasks, load balancing, and (potentially) back pressure.

The challenges presented above suggest the hardness of evaluating the performance of WSP systems. Indeed, they present all the typical challenges of stream processing engines with perceptibly more complex workloads. To aid the development of benchmarks and comparative studies, Scharrenbach et al. identify, in addition to the challenges, the following Key Performance Indicators [29]:

- Response time over the whole of the queries (Average/xth Percentile/Maximum).
- Maximum input throughput, i.e., the number of data items consumed per time unit.
- Minimum time to accuracy and minimum time to completion for all queries.

5.3.2 LSBench

In the absence of a standard benchmark, Le-Phuoc et al. [24] initially proposed a method and a framework to facilitate the comparisons of streaming linked data processing engines. Their evaluation framework, briefly named LSBench, includes a data generator to simulate realistic social media data (S2Gen) and 12 queries. Figure 5.6 presents LSBench logical schema for the benchmark data, which includes both static and streaming data from a social network use case. In particular, LSBench includes the following data streams:

- A GPS channel that reports the user location (latitude and longitude) over time.
- A Social Media channel that includes posts, likes, comments, and tags.
- A Multimedia Stream that includes photos and their associated attributes.

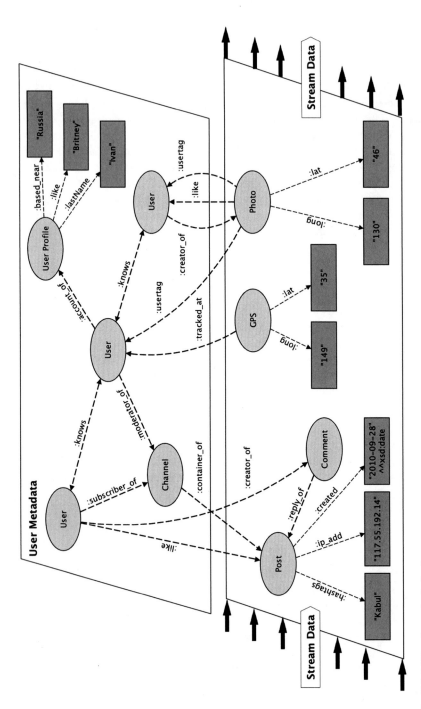

Fig. 5.6 LSBench ontological schema

Table 5.2 LSBench query comparison (Table from [24])

	Filter	Join	Aggregation	Nested	Negation	Union	TopK	Static data
Q1	✓							
Q2		✓						✓
Q3		✓						✓
Q4	✓	✓						
Q5		✓						✓
Q6		✓						✓
Q7	✓	✓					✓	✓
Q8		✓			✓			
Q9	✓	✓				✓		✓
Q10			✓					
Q11		✓	✓	✓				
Q12			✓				✓	

LSBench contains static data too, i.e., the user profile information and the relationships across the social network. In particular, every user owns a forum, and users can subscribe to each other's fora to read and reply to the posts and comments created there. Each forum is used as a channel for the posting stream of each user. LSBench covers three classes of testing: i.e., (a) query language expressiveness, (b) maximum execution throughput and scalability in terms of query number and static data size, and (c) result mismatch between different engines to assess the correctness of the answers.

LSBench includes 12 queries that aim at stressing the RSP engine under investigation. Table 5.2 shows a brief description of the benchmark queries according to the selected query features. Most of the queries have common SPARQL constructs, e.g., filters and BGPs with Joins. Notably, LSBench tests also the coverage and the performance of some advanced features like subqueries, Negation, and TopK (see Q11, Q8, and Q12, respectively). It is worth noticing that only Q2, Q3, Q5–Q7, and Q9 query both static and streaming data. Finally, no queries demand inference capabilities.

Last but not least, LSBench studies three Key Performance Indicators, i.e., (i) Feature Coverage, i.e., it tests if an engine supports a certain query (Boolean) (ii) Output Comparability, i.e., using a notion of mismatch, it tests if the engines are producing the same results (iii) Performance, i.e., it measures maximum input rate according to different scalability measures, e.g., size of static data or number of concurrent queries.

5.3.3   SRBench and CSRBench

The Streaming RDF/SPARQL benchmark and its extensions were the first benchmark designed specifically for RDF Stream Processing engines [36]. Figure 5.7 shows the different ontological schemas that the benchmark includes. SRBench incorporates data from the LinkedSensorWeb, mostly weather observations collected during hurricanes. The authors enriched such streams with static information from DBPedia[4] and GeoNames.[5] Zhang et al. focus on assessing the ability to deal with essential query features [36]. To this extent, SRBench defines 17 concise yet comprehensive queries that cover the major aspects of RSP tasks. The benchmark focuses on query language coverage, i.e., it verifies whether an engine supports the query. Zhang et al. accompany an analysis of the capabilities of $SPARQL_{stream}$, C-SPARQL Engine, and CQELS and their proposal query languages. Table 5.3 summarizes SRBench's queries in terms of features. In particular, it is worth mentioning that Q1–Q7 query only streaming data; Q8–Q11 query both streaming and background data; and Q12–Q17 additionally query external static data, i.e., GeoNames and DBpedia datasets. Moreover, Q3 and Q15–17 require RDFS inference capabilities.

CSRBench [15] is an extension of SRBench that takes into account some of the differences in the engine operation semantics (see Sect. 5.2.6). Dell'Aglio et al. noticed the different behavior of RSP engines with reference to time management. Therefore, they extended SRBench so to consider reporting policies, tick, windowing, whether or not empty results were reported, and input timestamping (i.e., the use of processing time). The result is that CSRBench adds the following to SRBench query set: (1) aggregated queries, (2) queries requiring to join triples with different timestamp, and (3) parametric sliding windows queries. Moreover, it provides an oracle to automatically check correctness of query results.

It is worth mentioning **YABench**, i.e., Yet Another RDF Stream Processing Benchmark [19], which focuses on evaluating both correctness and performance of the RSP engines. The benchmark uses the same weather observations as SRBench and CSRBench as use case and the same queries as CSRBench. Similar to CSRBench, YABench uses oracle-based validation but extends this notion to provide more comprehensive correctness metrics, e.g., precision and recall, for each window.

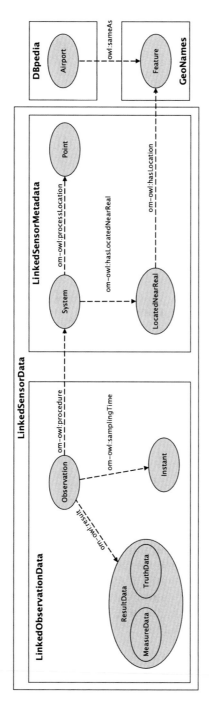

Fig. 5.7 SRBench ontological schema

Table 5.3 SRBench queries and their features

	Pattern matching	Solution modifier	Query form	SPARQL 1.1	Reasoning	Windows	Data access
Q1	A	P,D	S			✓	O
Q2	A,F,O	P,D	S	F,PP		✓	O
Q3	A	P	A	Ag	R	✓	O
Q4	A,F	P	S	Ag,E,M,F		✓	O
Q5	A	P	C	Ag,S		✓	O
Q6	A,F,U	P	S			✓	O
Q7	A	P,D	S	N		✓	O
Q8	A	P	S	Ag,E,M		✓	O, S
Q9	A	P	S	Ag,E,M		✓	O,S
Q10	A	P,D	S			✓	O,S
Q11	A,F	P,D	S	Ag,S,M,F		✓	O,S
Q12	A,F,U	P	S	Ag,S,E,M,F,PP		✓	O,S,F
Q13	A,F	P	S	Ag,E,M,F,PP		✓	O,S,G
Q14	A,F,U	P,D	S	F,PP		✓	O,S,G
Q15	A,F	P	S	Ag,E,M,PP	C	✓	O,S,D
Q16	A,F	P	S	PP	A		O,S,G,D
Q17	A,F	P	S	PP	C		S

Pattern matching: [A]nd, [F]ilter, [U]nion, [O]ptional. Solution modifiers: [P]rojection, [D]istinct, [S]elect, [C]onstruct, [A]sk, [Ag]gregate, [S]ubquery, [N]egation, [E]xpr in SELECT, assign[M]ent, [F]unctions, [P]operty [P]ath, sub[C]lassOf, subp[R]opertyOf, owl:same[A]s. Data access: Linked[O]servationData, Linked[S]ensorMetadata, [G]eoNames, [D]bpedia

5.3.4   CityBench

CityBench [1] provides real-world data streams from the CityPulse project[6] (i.e., vehicle traffic data, weather data, and parking spots data), and synthetic data streams about user location and air pollution. Real-world static datasets about cultural events are also included. Table 5.4 provides a brief description of the datasets, while Fig. 5.8 shows the ontological schema of CityBench data, which builds on the Semantic Sensor Network ontology [11].

The benchmark includes 13 continuous queries that are described in Table 5.5. The queries belong to different application scenarios in the smart city context. They are expressed in natural language and implemented in two RSP dialects, i.e., C-SPARQL and CQELS. They cover a number of advanced query features, e.g., advanced reporting (see Q5, Q7, and Q13), advanced aggregates (Q12), and TopK (Q9 and Q10). Several CityBench queries are parametric to allow scaling out experiments (e.g., Q1–Q3) or testing static selections (Q5, Q8) or stressing stream-to-stream joins (Q6).

---

[4] http://wiki.dbpedia.org/.

[5] http://www.geonames.org/ontology/documentation.html.

[6] http://citypulse.insight-centre.org/.

Table 5.4 CityBench dataset/datastreams description

Stream/dataset	Rate	Nature	Measures/descriptions
Traffic	5 min	Real-world	AVS, VC, ETT, CL
Parking	Unknown	Real-world	Number of vacant places
Sensor meta (SM)	Static	Real-world	SL, DSP, TR
Weather	Unknown	Real-world	DP, H, AP, T, WD, WS
Pollution	5 min	Synthesized	Air quality index
Cultural event (CE)	Quasi-static	Real-world	Cultural events provided by the municipality of Aarhus
Library events	Quasi-static	Real-world	A collection of 1548 library events hosted by libraries in the city
User location (UL)	Unknown	Synthesized	Geo-location coordinates of a fictional mobile user

Legend: [A]verage [V]ehicle [S]peed, [V]ehicle [C]ount, [E]stimated [T]ravel [T]ime, [C]ongestion [L]evel between the two points of road segment. [D]ew [P]oint (C), [H]umidity (%), [A]ir [P]ressure (mBar), [T]emperature (C), [W]ind [D]irection (○), and [W]ind [S]peed (kph). [S]ensor [L]ocation, [D]istance [S]ensors [P]airs, and [T]ype of [R]oads

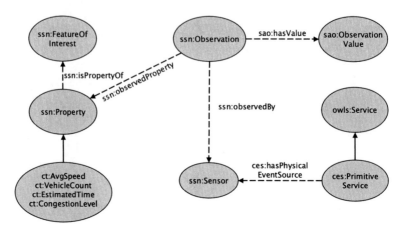

Fig. 5.8 CityBench data schema

CityBench allows to define a number of scalability parameters, i.e., (i) Number of Sensor Streams (for Queries Q1, Q2, Q3) (ii) Number of Concurrent Queries (iii) Background Data Size, i.e., several subsamples of cultural events and library events are provided. (iv) PlayBack Time, the data stream can be actualized to simplify querying in processing time. (v) Input Streaming Rate, the stream can be accelerated to stress-test the engines.

5.3.5   *Related Works*

The aforementioned benchmarks are specifically designed for evaluating Web Stream Processing engines. Indeed, they provide RDF datasets and streams and queries written in SPARQL-like languages and support reasoning. However, in the related literature, several engines were evaluated using alternative benchmarks, and the most relevant ones are briefly described below: (i) The **Linear road benchmark** [23] is a seminal benchmark in the context of stream processing systems. It was extensively used, extended, and adapted for WSP to evaluate systems like C-SPARQL engines (ii) **LUBM** [18] is probably the most popular reasoning benchmark, which features a scalable data generator, 12 queries requiring expressive reasoning. LUBM relates to WSP, thanks to the extensive works on incremental RDFS maintenance (iii) the **Berlin SPARQL benchmark** [8] is designed around an e-commerce use case with products, vendors and consumers, and reviews. The benchmark was used a few times to evaluate WSP systems, including CQELS [24] and Sparkwave [20] (iv) **SP^{2}Bench** [28] was also employed for CQELS evaluation [24]. The benchmark contains 11 queries and generates data based on DBLP RDF dataset.

Table 5.5 CityBench queries

Query	Projections	Selection	Agg	Modifier	Streams	Dataset	Reporting	Parameter
Q1	CL				Traffic	SM	Unspecified	Roads
Q2	CL, WC				Traffic, weather	SM	Unspecified	Roads
Q3	CL, ETT		AVG		Traffic	SM	Unspecified	Roads, destination
Q4			MIN		UL	CEs	Unspecified	
Q5	CL				Traffic	CEs	20 min window centered at event X	Cultural event X
Q6		1 km from current location			UL, parking	SM	Unspecified	Current location
Q7	Parking	Vacancies = 0			Traffic, parking	SM	On event	Destination
Q8	Parking	Vacancies > 0			Traffic, parking	CEs	Unspecified	Cultural event X
Q9	Parking, vacancies	Price = min	MIN	Order by	Parking	CEs	Unspecified	
Q10	AQ		MAX	Order by	Pollution	SM	On event	
Q11	Star	Count = 0	COUNT		Weather		10 min window	
Q12	Star	CL > Threshold	COUNT	HAVING	Traffic		20 min windows	Threshold
Q13	Star	CL > Threshold				On event	Threshold	

5.4 Methods and Tools

Alongside with the development of WSP engines and benchmarks, the Stream
Reasoning community also worked on defining methods and tools that facilitate
empirical research. Scharrenbach et al. initially distilled seven commandments for
the design and execution of WSP domain-specific benchmarks. Among existing
benchmarks [29], CityBench [1] and YABench [19] provide test drivers that should
simplify the development works. Last but not least, Tommasini et al. proposed
methods and tools for benchmarking window-based RSP engines [31, 32].

5.4.1 Heaven

Heaven is a methodology, proposed by Tommasini et al. [31], for evaluating
window-based RSP engines. It is based on two abstractions:

- Experiment, i.e., a test under controlled conditions that is made to examine the
  validity of a hypothesis.[7]
- Test Stand, i.e., a software facility that makes possible to design and execute
  experiments while collecting experimental measurements.

The abstractions are meant to guarantee *comparability* of the experimental results
and their relationship with the experimental conditions. Tommasini et al. define an
RSP Experiment as a tuple $< \mathcal{E}, \mathcal{T}, \mathcal{D}, \mathcal{Q}, \mathcal{K} >$ where [31]:

- $\mathcal{E}$ is the RSP engine used as subject in the experiment;
- $\mathcal{T}$ is an ontology and any data not subject to change during the experiment;
- $\mathcal{D}$ is the description of the input data streams;
- $\mathcal{Q}$ is the set of continuous queries registered into $\mathcal{E}$;
- $\mathcal{K}$ is the set of key performance indicators (KPIs) to collect.

The result of an RSP experiment is a report $\mathcal{R}$ that contains the traces of
the execution. According to Tommasini et al., the report $\mathcal{R}$ shall record (i) the
input stream with the timestamps, any KPIs to be measured before and after
processing, and, if any, the engine results from executing the queries. The test stand
operationalizes the experimental execution, creating an execution environment that
iterates over the experiment steps at a controlled pace and collecting performance
measurements in the meantime.

To guarantee *Reproducibility*, which refers to measurement variations on a
subject under changing conditions, the test stand must be engine-agnostic, i.e., any
RSP engine can be subject of an experiment. Moreover, the test stand shall be
data independent and *query independent*, allowing the usage of arbitrary streams,
datasets, and query from users' domains of interest. To guarantee *Repeatability*,

---

[7] https://www.yourdictionary.com/experiment.

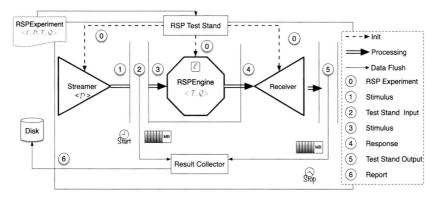

Fig. 5.9 Test stand logical and process views

which refers to variations on repeated measurements on a subject under identical conditions, the test stand shall minimize the experimental error, i.e., it has to affect the RSP engine evaluation as less as possible and in a predictable way.

Figure 5.9 presents the abstract architecture of a test stand that consists of four components:

- The STREAMER sets up the incoming streams w.r.t to $\mathcal{D}$;
- The RSP ENGINE represents the RSP engine $\mathcal{E}$, initialized with $\mathcal{T}$ and $\mathcal{Q}$;
- The RECEIVER continuously listens to $\mathcal{E}$ responses to $\mathcal{Q}$;
- The RESULT COLLECTOR compiles and persists the report $\mathcal{R}$.

Six kinds of events are exchanged between the components during the execution: (a) the top-level input, **RSP Experiment**, is used to initialize the test stand components (b) the incoming data streams in the form of **Stimuli** in which all the data have the same application time as specified in $\mathcal{D}$; (c) the answer of the queries specified in $\mathcal{Q}$, called **Response**; (d) the **TS Input** that combines every stimulus with the performance indicators measurements specified in $\mathcal{K}$ just before Stimulus creation; (e) the **TS Output** that combines every stimulus with the performance indicators measurements specified in $\mathcal{K}$ just before Response creation; (f) the **Report**—the carrier for the data that the Result Collector has to persist.

Figure 5.9 also shows how the test stand workflow is: the execution begins in (0), when the test stand receives an RSP Experiment, and it initializes its modules (dashed arrows): it sets up the engine $\mathcal{E}$ with $\mathcal{T}$ and $\mathcal{Q}$; it sets up the streamer, according to the incoming streams definition $\mathcal{D}$; it connects the engine to the Receiver and it initializes the Result Collector to receive and save the test stand outputs. Then, the test stand loops between steps (1) and (4), until the experiment ends.

The iterative execution proceeds as follows: at **(1)**, the Streamer creates and pushes a stimulus to the engine $\mathcal{E}$. At **(2)**, the test stand intercepts the stimulus of step (1), collects the performance measurements specified in $\mathcal{K}$, and sends a TS Input to the Result Collector. For instance, the test stand starts a timer to measure latency, and it calculates the memory consumption of the system. At **(3)**, the engine $\mathcal{E}$ receives a stimulus, and it processes it according with the query set Q. At **(4)**, which occurs whenever the engine $\mathcal{E}$ reports, a response is emitted, and at **(5)**, the Test Stand intercepts such response, takes the remaining measurements, and sends TS Output to the Result Collector. Finally, at **(6)**, the Result Collector persists all the collected data as a report.

5.4.2　CSRBench Oracle and YABench Driver

CRSBench and YAbench, in addition to the benchmark extension, provide a facility for measuring performance and result correctness. In particular, CRSBench includes an oracle for offline verification. Figure 5.10 shows the oracle architecture, which takes as input a stream S, a continuous query q, the specification of the engine operational semantics $< P, WC, EC, CC >$, and calculated system results R. The oracle checks the correctness of the query results by comparing it to the theoretical answer R.

YABench extends CSRBench oracle to a full-fledged test driver. In addition to correctness, YABench measures query result delivery, memory consumption, and CPU utilization. Figure 5.11 shows the test driver modular architecture: (I) the *Stream Generator* provisions the data streams. The rate of the stream is configurable and other generators could be plugged in, e.g., providing different scenarios. (II) The *RSP Engine* wraps the RSP engine under evaluation. Originally, it includes the C-SPARQL engine and CQELS, while additional engines can be plugged in. (III) The *Oracle* computes, for each window, precision and recall, delay of query results,

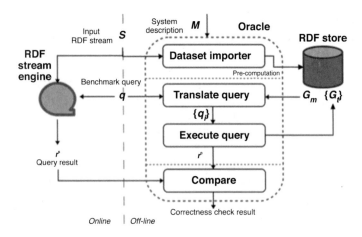

Fig. 5.10 RSP oracle from [15]

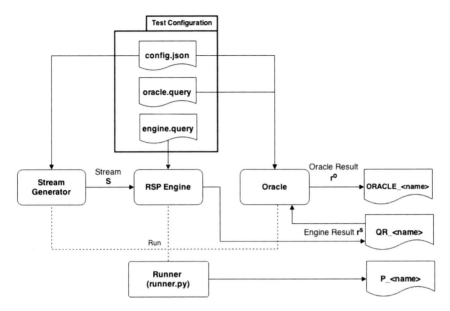

Fig. 5.11 YABench test driver from [19]

number of triples in the window, and number of tuples in the query results. The oracle can be configured basing on the engine *reporting policy*.

5.4.3   CityBench Testbed

CityBench provides a Testbed Infrastructure for experiment execution using CQELS and C-SPARQL engine that measures the KPI and stores the experiment results. The testbed is configurable, i.e., it allows choosing a variety of metrics including query latency, memory consumption, and result completeness. Figure 5.12 shows the testbed architecture, which includes three main modules:

- a **Dataset Configuration Module** for setting up stream-related metrics;
- a **Query Configuration Module** for configuring query-related metrics,
- a **Performance Evaluator**, which is responsible for recording and storing the measurements.

In practice, the testbed allows varying the experimental conditions. For instance, it is possible to perform changes in the input stream rate. The data generation can be tuned, for instance, to speed up or slow down the provided data streams. Moreover, the testbed performs data actualization. Indeed, it allows to customize the playback time of the data from any given time period to replay and mock up the exact situation during that period. Additionally, the testbed allows specifying which dataset to use as background knowledge, selecting from a provided set of datasets of different

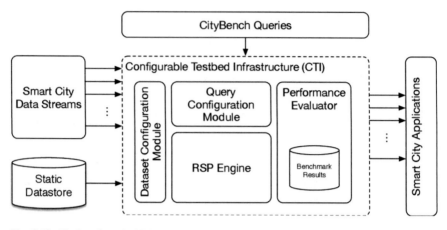

Fig. 5.12 CityBench testbed [1]

sizes. Finally, the CityBench testbed can measure the engine performance with reference to the number of concurrent queries and the number of streams within a given query. Users can specify any number of queries to be deployed for testing purposes. Similarly, to test the engine capability to deal with data distribution, the testbed allows configuring various sizes of streams involved within a single query.

5.4.4 RSPLab

RSPLab is a cloud-ready open-source test driver for evaluating window-based RSP engines [32]. It is based on the abstract architecture of Heaven's Test Stand, offers a programmatic environment design, and executes experiments. Figure 5.13 illustrates the RSPLab architecture that comprises four independent modules and sub-modules, and it mentions the technologies used for the current implementation, e.g., InfluxDB. The data provisioning tier, called *Streamer*, pushes RDF streams to the engine that is the subject of evaluation. The *Streamer* can provision any dataset that has a temporal dimension using a custom version of TripleWave [22]. The data processing tier, called *Consumer*, exposes the RSP engines on the Web by means of REST APIs [5]. Not all the RSP Services interfaces should be implemented; RSPLab requires source and query and sinks registration methods. The monitoring tier, called *Collector*, includes two sub-modules, i.e., a system for metric collection that continuously captures the performance statistics of any deployed module, and a storage module to persist the statistics and the query results. Finally, the control and analysis tier, simply called *Controller*, gives to the test driver users control on the other tiers. In practice, the controller allows to design and execute the experiments programmatically and enables results verification using an assisted and customized real-time data analysis dashboard.

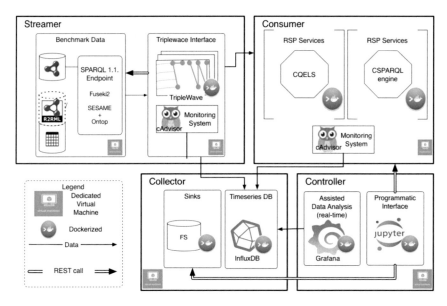

Fig. 5.13 RSPLab architecture [32]

5.4.5 Related Works

In addition to the WSP test driver presented above, LODLab, OLTP-Bench, and Gmark are worth mentioning:

LOD Lab [7] is a system from the Linked Data community that aims at reducing the human cost when evaluating triple stores. It supports data cleaning and simplifies dataset selections using metadata. Although LOD Lab can provision data using Web sockets, RSP engine evaluation is not in its scope. Indeed, it does not offer a continuous monitoring system nor continuous query answering.

OLTP-Bench [17] is a universal benchmarking infrastructure for relational databases. Similarly to RSPLab, it supports the deployment in a distributed environment, and it comes with assisted statistics visualization. However, it does not offer a programmatic environment to interact with the platform, execute experiments, or publish reports. OLTP-Bench includes a workload manager, but does not consider RDF streams. Moreover, it provides an SQL dialect translation module, which is flexible enough in the SQL area but not in the SR/RSP one.

Gmark is a framework for workload-centric triple store benchmarking. Gmark takes a schema-driven approach to the flexible and tightly controlled generation of synthetic graph instances. Although it supports sophisticated query workloads, it does not explicitly relates with RDF stream nor continuous query answering [4].

5.5   Chapter Summary

This chapter went through the most critical systems, benchmarks, and tools for Web Stream Processing. In particular, we focused on RDF Stream Processing engines like the prominent C-SPARQL and CQELS. We briefly discussed their architecture and their positioning in the literature. Moreover, we presented the existing domain-specific benchmarks for Web Stream Processing. Among others, CityBench and YABench are the most recent and complete ones. Finally, we discussed the methods and tools that the stream reasoning community designed to reduce the experimental efforts. Heaven, RSPLab, and the CityBench configurable testbed are essential steps to reduce the exhausting work that precedes the performance assessment of new techniques for Web stream data management.

References

1. Ali, Muhammad Intizar, Feng Gao, and Alessandra Mileo. 2015. Citybench: A configurable benchmark to evaluate RSP engines using smart city datasets. In *Proceedings of The Semantic Web – ISWC 2015 – 14th International Semantic Web Conference, Bethlehem, October 11–15, 2015, Part II*, 374–389.
2. Allen, James F. 1991. Time and time again: The many ways to represent time. *International Journal of Intelligence Systems* 6(4): 341–355.
3. Anicic, Darko, Paul Fodor, Sebastian Rudolph, and Nenad Stojanovic. 2011. EP-SPARQL: a unified language for event processing and stream reasoning. In *Proceedings of the 20th International Conference on World Wide Web, WWW 2011, Hyderabad, March 28–April 1, 2011*, ed. Sadagopan Srinivasan, Krithi Ramamritham, Arun Kumar, M. P. Ravindra, Elisa Bertino, and Ravi Kumar, 635–644. New York: ACM.
4. Bagan, Guillaume, Angela Bonifati, Radu Ciucanu, George H. L. Fletcher, Aurélien Lemay, and Nicky Advokaat. 2017. gMark: Schema-driven generation of graphs and queries. *IEEE Transactions on Knowledge and Data Engineering* 29 (4): 856–869
5. Balduini, Marco, and Emanuele Della Valle. 2013. A restful interface for RDF stream processors. In *International Semantic Web Conference (Posters & Demos)*. Vol. 1035. *CEUR Workshop Proceedings*, 209–212. CEUR-WS.org.
6. Barbieri, Davide Francesco, Daniele Braga, Stefano Ceri, Emanuele Della Valle, and Michael Grossniklaus. 2010. C-SPARQL: a continuous query language for RDF data streams. *International Journal of Semantic Computing* 4 (1): 3–25.
7. Beek, Wouter, Laurens Rietveld, Filip Ilievski, and Stefan Schlobach. 2016. LOD lab: Scalable linked data processing. In *Reasoning Web*. Vol. 9885. *Lecture Notes in Computer Science*, 124–155. Berlin: Springer.
8. Bizer, Christian, and Andreas Schultz. 2011. The berlin SPARQL benchmark. In *Semantic Services, Interoperability and Web Applications – Emerging Concepts*, ed. Amit P. Sheth, 81–103. Boca Raton: CRC Press.
9. Botan, Irina, Roozbeh Derakhshan, Nihal Dindar, Laura M. Haas, Renée J. Miller, and Nesime Tatbul. 2010. SECRET: A model for analysis of the execution semantics of stream processing systems. *PVLDB* 3 (1): 232–243.
10. Calbimonte, Jean-Paul, Hoyoung Jeung, Oscar Corcho, and Karl Aberer. 2012. Enabling query technologies for the semantic sensor web. *International Journal On Semantic Web and Information Systems (IJSWIS)* 8 (1): 43–63.

11. Compton, Michael, Payam M. Barnaghi, Luis Bermudez, Raul Garcia-Castro, Óscar Corcho, Simon J. D. Cox, John Graybeal, Manfred Hauswirth, Cory A. Henson, Arthur Herzog, Vincent A. Huang, Krzysztof Janowicz, W. David Kelsey, Danh Le Phuoc, Laurent Lefort, Myriam Leggieri, Holger Neuhaus, Andriy Nikolov, Kevin R. Page, Alexandre Passant, Amit P. Sheth, and Kerry Taylor. 2012. The SSN ontology of the W3C semantic sensor network incubator group. *Journal of Web Semantics* 17: 25–32.

12. Dell'Aglio, Daniele, and Emanuele Della Valle. 2014. Incremental reasoning on RDF streams. In *Linked Data Management*, 413–435. London: Chapman and Hall/CRC.

13. Dell'Aglio, Daniele, Emanuele Della Valle, Jean-Paul Calbimonte, and Óscar Corcho. 2014. RSP-QL semantics: A unifying query model to explain heterogeneity of RDF stream processing systems. *International Journal on Semantic Web and Information Systems* 10 (4).

14. Dell'Aglio, Daniele, Emanuele Della Valle, Frank van Harmelen, and Abraham Bernstein. 2017. Stream reasoning: A survey and outlook. *Data Science* 1 (1–2): 59–83.

15. Dell'Aglio, Daniele, Jean-Paul Calbimonte, Marco Balduini, Óscar Corcho, and Emanuele Della Valle. 2013. On correctness in RDF stream processor benchmarking. In *International Semantic Web Conference (2)*. Vol. 8219. *Lecture Notes in Computer Science*, 326–342. Berlin: Springer.

16. Dell'Aglio, Daniele, Minh Dao-Tran, Jean-Paul Calbimonte, Danh Le Phuoc, and Emanuele Della Valle. 2016. A query model to capture event pattern matching in RDF stream processing query languages. In *EKAW*. Vol. 10024. *Lecture Notes in Computer Science*, 145–162.

17. Difallah, Djellel Eddine, Andrew Pavlo, Carlo Curino, and Philippe Cudré-Mauroux. 2013. OLTP-Bench: An extensible testbed for benchmarking relational databases. *PVLDB* 7 (4): 277–288.

18. Guo, Yuanbo, Zhengxiang Pan, and Jeff Heflin. 2005. LUBM: A benchmark for OWL knowledge base systems. *Journal of Web Semantics* 3 (2–3): 158–182.

19. Kolchin, Maxim, Peter Wetz, Elmar Kiesling, and A Min Tjoa. 2016. Yabench: A comprehensive framework for RDF stream processor correctness and performance assessment. In *Proceedings of the Web Engineering – 16th International Conference, ICWE 2016, Lugano, June 6–9, 2016*, 280–298.

20. Komazec, Srdjan, Davide Cerri, and Dieter Fensel. 2012. Sparkwave: Continuous schema-enhanced pattern matching over RDF data streams. In *DEBS*, 58–68. New York: ACM.

21. Le-Phuoc, Danh, Minh Dao-Tran, Josiane Xavier Parreira, and Manfred Hauswirth. 2011. A native and adaptive approach for unified processing of linked streams and linked data. In *International Semantic Web Conference*, 370–388. Berlin: Springer.

22. Mauri, Andrea, Jean-Paul Calbimonte, Daniele Dell'Aglio, Marco Balduini, Marco Brambilla, Emanuele Della Valle, and Karl Aberer. 2016. Triplewave: Spreading RDF streams on the web. In *ISWC*.

23. Michael, Panayiotis Adamos, and Douglas Stott Parker Jr. 2008. Architectural principles of the "streamonas" data stream management system and performance evaluation based on the linear road benchmark. In *International Conference on Computer Science and Software Engineering, CSSE 2008*. Vol. 4, *Embedded Programming/Database Technology/Neural Networks and Applications/Other Applications, December 12–14, 2008, Wuhan*, 643–646. Washington: IEEE Computer Society.

24. Phuoc, Danh Le, Minh Dao-Tran, Minh-Duc Pham, Peter A. Boncz, Thomas Eiter, and Michael Fink. 2012. Linked stream data processing engines: Facts and figures. In *International Semantic Web Conference (2)*. Vol. 7650. *Lecture Notes in Computer Science*, 300–312. Berlin: Springer.

25. Ren, Xiangnan, and Olivier Curé. 2017. Strider: A hybrid adaptive distributed rdf stream processing engine. In *International Semantic Web Conference*, 559–576. Berlin: Springer.

26. Rinne, Mikko, Esko Nuutila, and Seppo Törmä. 2012. INSTANS: High-performance event processing with standard RDF and SPARQL. In *11th International Semantic Web Conference ISWC*. Vol. 914, 101–104. Citeseer.

27. Rinne, Mikko, Esko Nuutila, and Seppo Törmä. 2012. INSTANS: high-performance event processing with standard RDF and SPARQL. In *Proceedings of the ISWC 2012 Posters & Demonstrations Track, Boston, November 11–15, 2012*, ed. Birte Glimm and David Huynh. Vol. 914. *CEUR Workshop Proceedings*. CEUR-WS.org.

28. Schmidt, Michael, Thomas Hornung, Michael Meier, Christoph Pinkel, and Georg Lausen. 2009. Sp^2bench: A SPARQL performance benchmark. In *Semantic Web Information Management - A Model-Based Perspective*, ed. Roberto De Virgilio, Fausto Giunchiglia, and Letizia Tanca, 371–393. Berlin: Springer.

29. Thomas Scharrenbach, Jacopo Urbani, Alessandro Margara, Emanuele Della Valle, and Abraham Bernstein. 2013. Seven commandments for benchmarking semantic flow processing systems. In *Proceedings of The Semantic Web: Semantics and Big Data, 10th International Conference, ESWC 2013, Montpellier, May 26–30, 2013*, 305–319.

30. Tommasini, Riccardo, Pieter Bonte, Emanuele Della Valle, Erik Mannens, Filip De Turck, and Femke Ongenae. 2016. Towards ontology-based event processing. In *OWLED*. Vol. 10161. *Lecture Notes in Computer Science*, 115–127. Berlin: Springer.

31. Tommasini, Riccardo, Emanuele Della Valle, Marco Balduini, and Daniele Dell'Aglio. 2016. Heaven: A framework for systematic comparative research approach for RSP engines. In *ESWC*. Vol. 9678. *Lecture Notes in Computer Science*, 250–265. Berlin: Springer.

32. Tommasini, Riccardo, Emanuele Della Valle, Andrea Mauri, and Marco Brambilla. 2017. RSPLab: RDF stream processing benchmarking made easy. In *ISWC*, 202–209.

33. Tommasini, Riccardo, Pieter Bonte, Femke Ongenae, and Emanuele Della Valle. 2021. RSP4J: an API for RDF stream processing. In *Proceedings of The Semantic Web – 18th International Conference, ESWC 2021, Virtual Event, June 6–10, 2021*, ed. Ruben Verborgh, Katja Hose, Heiko Paulheim, Pierre-Antoine Champin, Maria Maleshkova, Óscar Corcho, Petar Ristoski, and Mehwish Alam. Vol. 12731. *Lecture Notes in Computer Science*, 565–581. Berlin: Springer.

34. Walavalkar, Onkar, Anupam Joshi, Tim Finin, Yelena Yesha, et al. 2008. Streaming knowledge bases. In *Proceedings of the Fourth International Workshop on Scalable Semantic Web knowledge Base Systems*.

35. White, Walker M., Mirek Riedewald, Johannes Gehrke, and Alan J. Demers. 2007. What is "next" in event processing? In *Proceedings of the Twenty-Sixth ACM SIGACT-SIGMOD-SIGART Symposium on Principles of Database Systems, June 11–13, 2007, Beijing*, ed. Leonid Libkin, 263–272. New York: ACM.

36. Zhang, Ying, Minh-Duc Pham, Óscar Corcho, and Jean-Paul Calbimonte. 2012. Srbench: A streaming RDF/SPARQL benchmark. In *Proceedings of The Semantic Web – ISWC 2012 – 11th International Semantic Web Conference, Boston, November 11–15, 2012, Part I*, 641–657.

Chapter 6
Exercise Book

Abstract This chapter introduces hands-on practical examples and exercises in RSP4J. It first details a location-detection example and investigates how queries can be defined in RSP-QL and how the underlying operators can be specified directly in RSP4J. At the time of writing, COVID-19 is still one of the most discussed topics. This chapter introduces a number of practical examples around contact tracing and detection of potentially infected individuals. Next to the location-detection example, this chapter also presents illustrative examples of Web streams. In particular, this chapter features `DBPedia Live` (https://wiki. dbpedia.org/online-access/DBpediaLive) that shares the RDF changes on DBPedia, `Wikimedia EventStreams` (https://stream.wikimedia.org) that streams the changes across all the WikiMedia projects, and the `Global Database of Events, Languages, and Tone (GDELT)` (https://gdeltproject.org) streams.

6.1 Introduction

This chapter introduces hands-on practical examples and exercises in RSP4J [5]. It first details a location-detection example and investigates how queries can be defined in RSP-QL [1] and how the underlying operators can be specified directly in RSP4J. At the time of writing, COVID-19 is still one of the most discussed topics. Section 6.2 introduces a number of practical examples around contact tracing and detection of potentially infected individuals. Next to the location-detection example, this chapter presents illustrative examples of Web streams. Section 6.3 explains the comprehensive set of challenges that need to be tackled when the life cycle is applied to real-world data. In particular, this chapter features `DBPedia Live`[1] that shares the RDF changes on DBPedia, `Wikimedia EventStreams`[2] that streams the

---

[1] https://wiki.dbpedia.org/online-access/DBpediaLive.

[2] https://stream.wikimedia.org.

© The Author(s), under exclusive license to Springer Nature Switzerland AG 2023
R. Tommasini et al., *Streaming Linked Data*,
https://doi.org/10.1007/978-3-031-15371-6_6

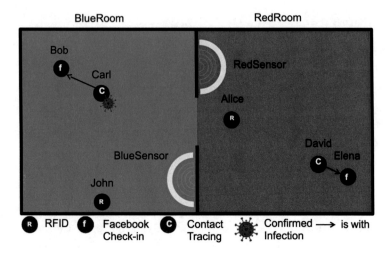

Fig. 6.1 The COVID-19 scenario setup: two adjacent rooms, six persons, three streams of possible location updates, and a stream of COVID test results

changes across all the WikiMedia projects, and the `Global Database of Events, Languages, and Tone (GDELT)`[3] streams.

6.2   Location-Detection Example

6.2.1   The Setup

In the COVID-19 scenario, a simplified setup consisting of two adjacent rooms will be used, i.e., a Blue Room and a Red Room, as shown in Fig. 6.1. There are several people in each room; however, they can move from one room to another at any time. To know the location of each person, there are three localization options:

1. *RFID*: several persons are equipped with an RFID tag, and each room has its own RFID sensor, i.e., BlueSensor and RedSensor. When a person with an RFID tag is near one of the sensors, it is captured by the sensors and communicated on an *Observation Stream*.
2. *Facebook Check-ins*: another way to capture one's locations is by monitoring the Check-ins certain individuals have done on Facebook. When certain persons have checked in into the BlueRoom on Facebook, it can be assumed that they will be present in that room for a certain amount of time. These location updates will also be communicated in the *Observation Stream*.

---

[3] https://gdeltproject.org.

3. *Contact Tracing*: localization through contact tracing is a bit different than the previous two examples, as it only allows to capture persons' relative locations by proximity to another individual. These kinds of updates thus tell that two individuals were close to each other during a certain amount of time, but not where they were exactly. The proximity of individuals will be communicated in the *Contact Stream*.

Next to the location updates, the COVID-19 test results will also be captured in a *COVID Stream*. This stream describes the test results in terms of COVID-19 infections.

6.2.2 The Data

Now that the scenario has been described and a closer look has been taken on the four streams that will be processed, let us zoom in on how the data is described in each of these streams. Figure 6.2 visualizes the data in the different streams both in tabular form as in RDF. Note that in a realistic scenario, chances are high that each of these different streams uses different formats, e.g., JSON, XML, CSV, etc. Using RDF is ideal in this case as it allows to integrate data from different sources. For the sake of clarity, the data is only described in tabular format.

1. *RFID*: the RFID updates detail the sensor that made the observation, the room the sensor is located in, and the person who was detected.
2. *Facebook Check-ins*: the Facebook Check-ins detail the person that did the check-in and the room in which they checked in. Note that in RDF, the check-ins are very similar to the *RFID observations*. For simplicity, both are combined in

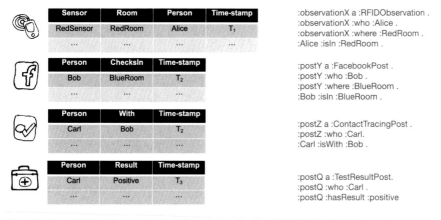

Fig. 6.2 The COVID-19 scenario data: the content of each stream is detailed both in tabular as in RDF format

the same stream, i.e., the *Observation Stream*. Note that the *Observation Stream* can thus contain both *RFID Observations* and *Facebook Posts*.

3. *Contact Tracing*: these updates describe who is with who, but not in which room they are located. These updates will be on the *Contact Stream*.

4. *Confirmed Infections*: the confirmed infections detail the person who did the test and the test results, i.e., either positive or negative. These updates will be on the *COVID Stream*.

6.2.3 The Queries

Now that the data in the different streams is known, let's take a look at the queries that will be answered:

- *Query 1: who is with who and in which room?* The first query detects which persons are nearby and figures out in which room they are. The *RFID, Facebook Check-ins*, and *Contact Tracing* streams will be used. In Fig. 6.1, *Query 1* would indicate that *Dave and Elena are close together and are in the Red Room*.
- *Query 2: who might be infected through close contact?* The second query detects who might be infected because they were in close contact with someone who had a positive test result. The *RFID, Facebook Check-ins, Contact Tracing*, and *Confirmed Infections* streams will be used.
- *Query 3: who is in a room containing an infected person?* This query retrieves the persons that were inside a room that happen to contain an infected person. It combines the *RFID, Facebook Check-ins, Contact Tracing*, and *Confirmed Infections* streams.

6.2.4 Solving Query 1

For all queries, solutions will be provided both using RSP-QL query syntax and using RSP4J that defines the different operators separately. The latter allows us to understand what is going on under the hood. Try to write the query on paper, or you can try it directly out using the accompanying examples on GitHub.[4] The code on GitHub has a generator that populates the streams, allowing you to test out your solution.

Query 1 RSP-QL Solution

To solve the first query, two streams will be used: the *ObservationStream* containing both the *RFIDObservations* and *Facebook Posts* and the *ContactStream* containing

---

[4] https://github.com/StreamingLinkedData/Book.

the *Contact Tracing* results. In terms of processing steps as defined in Sect. 4.3.8, a *Filter* is necessary to extract the statements detailing the location of the person in the *ObservationStream* and who is with whom in the *ContactStream*. After the *Filter*, a *Merge* is necessary to combine the two streams together. The *Observation-Stream* is available at *http://rsp4j.io/covid/observations* and the *ContactStream* at *http://rsp4j.io/covid/tracing*. Each of the streams needs a definition of a window. For this query, a window of 10 minutes is defined which allows to join the two streams together. Listing 6.1 shows the solution of *Query 1*.

```
 1 PREFIX : <http://rsp4j.io/covid/>
 2 PREFIX rsp4j: <http://rsp4j.io/>
 3 REGISTER RSTREAM <http://out/stream> AS
 4 SELECT ?person1 ?room ?person2
 5 FROM NAMED WINDOW rsp4j:window ON :observations [RANGE PT10M STEP PT1M]
 6 FROM NAMED WINDOW rsp4j:window2 ON :tracing [RANGE PT10M STEP PT1M]
 7 WHERE {
 8   WINDOW rsp4j:window { ?person1 :isIn ?room .}
 9   WINDOW rsp4j:window2 { ?person2 :isWith ?person1 .}
10 }
11
```

Listing 6.1 The RSP-QL query that determines who is with whom and in what room

Lines 1–2 defines the prefixes, line 3 registers the output of the resulting stream as an *RStream*, line 4 defines the *Query Form*, and lines 5–6 defines the windows on the two streams. Line 8–9 queries the content of each of the windows (i.e., *Filter* operation) and joins them together based on the variable *?person1* that is present in both windows (i.e. *Merge* operation). This simple query allows us to solve *Query 1* and tells us *Who is with Who and in which room*.

Query 1 Operator Solution

Now that RSP-QL that can solve *Query 1* has been discussed, let us try to define the operators manually, through RSP4J. First, Listing 6.2 defines the engine properties, such as the reporting policies and the tick.

```
// Engine properties
Report report = new ReportImpl();
report.add(new OnWindowClose());
Tick tick = Tick.TIME_DRIVEN;
ReportGrain report_grain = ReportGrain.SINGLE;
Time instance = new TimeImpl(0);

RSPEngine engine = new RSPEngine(instance, tick, report_grain,
    report);
```

Listing 6.2 Solving Query 1 with RSP4J: engine definition via abstraction API

Based on these engine properties, the two windows can be manually defined, i.e., one for the observation stream and one for the tracing stream, as shown in Listing 6.3.

```
// Window (S2R) declaration
StreamToRelationOp<Graph, Graph> w1 = engine.createCSparqlWindow(
RDFUtils.createIRI("window1"),
10*60*1000, // window width in milliseconds
60*1000);   // window  slide in milliseconds

StreamToRelationOp<Graph, Graph> w2 = engine.createCSparqlWindow(
RDFUtils.createIRI("window2"),
10*60*1000, // window width in milliseconds
60*1000);   // window slide in milliseconds
```

Listing 6.3 Solving Query 1 with RSP4J: window definition via abstraction API

Next, Listing 6.4 defines the R2R operator through two BGPs.

```
// Definition of the prefixes
PrefixMap prefixes = new PrefixMap();
prefixes.addPrefix("","http://rsp4j.io/covid/");

// Definition of the R2R operators
// BGP for window 1
BGP bgp = BGP.createWithPrefixes(prefixes)
.addTP("?person1", ":isIn", "?room")
.build();
// BGP for window 2
BGP bgp2 = BGP.createWithPrefixes(prefixes)
.addTP("?person2", ":isWith", "?person1")
.build();
```

Listing 6.4 Solving Query 1 with RSP4J: R2R definition via abstraction API

Finally, Listing 6.5 combines everything together in order to create a *Task* with similar properties as the RSP-QL query defined earlier.

```
// Create the RSP4J Task and Continuous Program
TaskAbstractionImpl<Graph, Graph, Binding, Binding> t =
  new TaskAbstractionImpl.TaskBuilder(prefixes)
    .addS2R(":observations", w1, "window1")
    .addS2R(":tracing", w2, "window2")
    .addR2R("window1", bgp)
    .addR2R("window2", bgp2)
    .addR2S("out", new Rstream<Binding, Binding>())
    .addProjectionStrings(List.of("?s","?o","?s2"))
    .build();
ContinuousProgram<Graph, Graph, Binding, Binding> cp =
  new ContinuousProgram.ContinuousProgramBuilder()
    .in(observationStream)
    .in(tracingStream)
    .in(covidStream)
    .addTask(t)
    .out(outStream)
    .addJoinAlgorithm(new HashJoinAlgorithm())
    .build();
```

Listing 6.5 Solving Query 1 with RSP4J: engine definition via abstraction API

Looking under the hook gives us a clearer view of what is happening inside our
RSP engine.

6.2.5   Solving Query 2

Now the query *Who might be infected through close contact?* will be solved. This
query is a little more complex, as it needs to combine three streams. Furthermore,
looking at the results of the contact tracing, which states that a certain person is
nearby of another person, either one of these persons can be infected. Thus, when
combining the stream with the test results, this consideration needs to be taken into
account.

Query 2 RSP-QL Solution

To solve the second query, the following three streams will be used: the *ObservationStream* containing both the *RFIDObservations* and *Facebook Posts*, the *ContactStream* containing the *Contact Tracing* results, and the *CovidStream* describing
the COVID test results. The *ObservationStream* is available at *http://rsp4j.io/covid/
observations*, the *ContactStream* is available at *http://rsp4j.io/covid/tracing*, and the
CovidStream is available at *http://rsp4j.io/covid/testResults*.

On each of the streams, a window needs to be defined, and a window of 10
minutes is chosen to join the observation and tracing stream together. However, it
can be assumed that the COVID test results are valid for 1 day, i.e., 24 hours. As in
the first query, first the data needs to be *Filtered* data, and then the filtered streams
need to be *Merged* together. Listing 6.6 shows the solution of *Query 2*.

```
 1  PREFIX : <http://rsp4j.io/covid/>
 2  PREFIX rsp4j: <http://rsp4j.io/>
 3  REGISTER RSTREAM <http://out/stream> AS
 4  SELECT ?person1 ?room ?person2 ?person3
 5  FROM NAMED WINDOW rsp4j:window1 ON :observations [RANGE PT10M STEP PT1M]
 6  FROM NAMED WINDOW rsp4j:window2 ON :tracing [RANGE PT10M STEP PT1M]
 7  FROM NAMED WINDOW rsp4j:window3 ON :testResults [RANGE PT24H STEP PT1M]
 8  WHERE {
 9      WINDOW rsp4j:window1 { ?person1 :isIn ?room .}
10      WINDOW rsp4j:window2 { ?person2 :isWith ?person1 .}
11      WINDOW rsp4j:window3 { ?testResult a :TestResultPost;
12                                    :who ?person3;
13                                    :hasResult :positive .}
14      FILTER(?person3 = ?person1 || ?person3 = ?person2).
15      }
```

Listing 6.6 The RSP-QL query that determines who is with whom and in what room

This query is a little more complex, as the test results can be for either person1 or person2, reported through the contact tracing application. As it is unknown in advance, which person could be infected, an additional *FILTER* clause is added that checks if either of the involved persons were indeed infected. In Listing 6.6, line 12 creates a new variable for the person that has a positive test result. Line 14 then checks if either person1 or person2 is the person which received the positive result. Note that line 7 has defined the window on top of the test results of the last 24 hours. In practice, the results would be valid for a couple of days, but this example shows that different sizes of windows can be defined in the same query.

Query 2 Operator Solution

The solution of Query 2 through the query operators will be left as an exercise. The solution can be found on the GitHub page.[5]

6.2.6   Solving Query 3

Now the query *Who is in a room containing an infected person?* will be solved, which requires the combination of all three streams. This problem is a little more complex, as some persons are not explicitly located in a certain room, however only implicitly through contact tracing. For example, in Fig. 6.1, Elena is located in the red room and David is with Elena. Even though the location of David is not explicitly stated, it can be inferred that David is also in the red room. Query

---

[5] https://github.com/StreamingLinkedData/Book.

definition could become rather difficult if it needs to check for each person if they are with another person for whom the location is known. Note this chain can grow, e.g., Carl could be detected with David, while the latter is located with Elena. Through the use of the processing step *Lifting*, implicit data can be inferred. The following rule can be defined, stating that when a person is with another person and the latter person's location is known, then the former person has the same location:

(R1) ?p :isIn ?room ← ?p2 :isIn ?room AND ?p :isWith ?p2

Adding this rule allows to query the location of each person, even though they were only implicitly located through contact tracing.

Query 3 Operator Solution

To solve the third query, three streams will be used: the *ObservationStream* containing both the *RFIDObservations* and *Facebook Posts*, the *ContactStream* containing the *Contact Tracing* results, and the *CovidStream* describing the COVID test results. The *ObservationStream* is available at *http://rsp4j.io/covid/observations*, the *ContactStream* is available at *http://rsp4j.io/covid/tracing*, and the *CovidStream* is available at *http://rsp4j.io/covid/testResults*.

On each of the streams, a window needs to be defined, and a window of 10 minutes is chosen to join the observation and tracing stream together. However, it is again assumed that the COVID test results are valid for 1 day, i.e., 24 hours. As in the first and second query, it first needs to *Filter* the data and then *Merge* the filtered streams together. For this third query, a *Lift* processing operation is required to retrieve the implicit locations of each person. This is done through the rule defined above and should be loaded into the engine. Also note that when evaluating the rule, the window on the *ObservationStream* and *ContactStream* needs to be combined.

In order to define the operators manually, Listing 6.7 defines the windows on each stream and reuses the engine properties from Listing 6.2.

```
// Window (S2R) declaration
StreamToRelationOp<Graph, Graph> w1 = engine.createCSparqlWindow(
RDFUtils.createIRI("window1"),
    10*60*1000, // window width in milliseconds
    60*1000);   // window  slide in milliseconds

StreamToRelationOp<Graph, Graph> w2 = engine.createCSparqlWindow(
RDFUtils.createIRI("window2"),
    10*60*1000, // window width in milliseconds
    60*1000);   // window slide in milliseconds

StreamToRelationOp<Graph, Graph> w3 = engine.createCSparqlWindow(
RDFUtils.createIRI("window3"),
    24*60*60*1000, // window width in milliseconds
    60*1000);   // window slide in milliseconds
```

Listing 6.7 Solving Query 3 with RSP4J: window definition via abstraction API

Next, in Listing 6.8, the R2R operators are defined. Note that the *Observation-Stream* and *ContactStream* can be combined in order to execute the predefined rule. Note that the second BGP is much more simple compared to the second query, as it does not need to check if infected persons were implicitly or explicitly located.

```
// Definition of the R2R operators
// BGP for window 1 and 2
BGP bgp = BGP.createWithPrefixes(prefixes)
.addTP("?person", ":isIn", "?room")
.build();
// BGP for window 3
BGP bgp2 = BGP.createWithPrefixes(prefixes)
.addTP("?testResult", "a", ":TestResultPost")
.addTP("?testResult",":who", "?person")
.addTP("?testResult",":hasResult",":positive")
.build();
```

Listing 6.8 Solving Query 3 with RSP4J: R2R operator via abstraction API

In order to define the *Lift* and the execution of the predefined rule, an *R2RPipe* can be created that links the evaluation of the rule with the evaluation of the BGP, as defined in Listing 6.9. Also note that the *DatalogR2R* is able to evaluate datalog rules such as the one defined above.

```
// Define a rule that infers the location of individuals reported
    through the contact tracing stream
DatalogR2R datalogR2R = new DatalogR2R();
TripleGenerator tripleGenerator = new TripleGenerator(prefixes);

ReasonerTriple head = tripleGenerator
    .createReasonerTriple("?p", ":isIn", "?room");
ReasonerTriple body1 = tripleGenerator
    .createReasonerTriple("?p", ":isWith", "?p2");
ReasonerTriple body2 = tripleGenerator
    .createReasonerTriple("?p2", ":isIn", "?room");

Rule r = new Rule(head,body1,body2);
datalogR2R.addRule(r);
// Create a pipe of two r2r operators, datalog reasoner and BGP
R2RPipe<Graph,Binding> r2r = new R2RPipe<>(datalogR2R,bgp);
```

Listing 6.9 Solving Query 3 with RSP4J: lifting operator via abstraction API

Finally, Listing 6.10 combines everything together. Note the evaluation of the R2RPipe, combining the rule and bgp over both window1 and window2.

```
// Create the RSP4J Task and Continuous Program
TaskOperatorAPIImpl<Graph, Graph, Binding, Binding> t =
        new TaskOperatorAPIImpl.TaskBuilder(prefixes)
                .addS2R(":observations", w1, "window1")
                .addS2R(":tracing", w2, "window2")
                .addS2R(":testResults", w3, "window3")
                .addR2R(List.of("w1", "w2"), r2r)
                .addR2R("w3", bgp2)
                .addR2S("out", new Rstream<Binding, Binding>())
                .build();
ContinuousProgram<Graph, Graph, Binding, Binding> cp =
        new ContinuousProgram.ContinuousProgramBuilder()
                .in(observationStream)
                .in(tracingStream)
                .in(covidStream)
                .addTask(t)
                .out(outStream)
                .addJoinAlgorithm(new HashJoinAlgorithm())
                .build();
```

Listing 6.10 Solving Query 3 with RSP4J: engine definition via abstraction API

> **Query 3 RSP-QL Solution**

The solution of Query 3 through RSP-QL is left as an exercise. The solution can be found on the GitHub page.[6]

6.3   Publishing and Processing Wild Streams

Now it is time to focus on three representative examples of real-world Web Streams published as Streaming Linked Data, i.e., DBPedia Live, Wikimedia EventStream, and the Global Dataset of Event, Language, and Tone (GDELT). Our goal is to provide examples of how to publish real-world Web Streams and show how they can be processed using RSP-QL. Figure 6.3 positions each of real-world stream w.r.t. the publishing scenarios from Fig. 4.1.

---

[6] https://github.com/StreamingLinkedData/Book.

Fig. 6.3 Positioning of
real-world streams w.r.t.
publishing scenarios

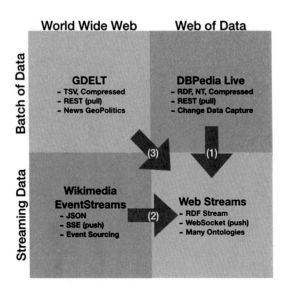

6.3.1   DBPedia Live

DBPedia is a community project that extracts structured data from Wikipedia in the form of an Open Knowledge Graph. DBPedia is served as linked data using Entity-Pages, REST APIs, or SPARQL endpoints. The latest version of DBPedia counts more than 470 million of RDF triples.

DBPedia Live (DBL)[7] is a changelog stream continuously published to keep DBPedia replicas in synch. A synchronization tool designed to consume the batches and update the local DBPedia copy is available [3]. DBL uses a pull-based mechanism for data provisioning. It shares RDF data using DBPpedia ontology (DBO), which is a cross-domain ontology, manually created from the most used Wikipedia info-boxes. The latest version of DBO covers 685 classes and 2795 different properties. It is a direct-acyclic graph, i.e., classes may have multiple superclasses, as required to map it to schema.org.

```
:dbl a vocals:StreamDescriptor ;
dcat:title "DPedia Live" ; dcat:publisher <http://www.dbpedia.org> ;
rdfs:seeAlso <https://wiki.dbpedia.org/services-resources/ontology>
dcat:license <https://creativecommons.org/licenses/by-nc/4.0/> ;
dcat:publisher <http://www.linkeddata.stream/about> ;
dcat:dataset :dblstream   .
:dblstream a vocals:RDFStream ;
vocals:hasEndpoint :dblendpoint, :dblendpointold .
```

Listing 6.11   A DBL Stream VoCaLS description. Prefixes omitted

---

[7] http://dbpedia-live.openlinksw.com/live/.

A DBL update consists of four compressed N-Triples (NT) files. Two main different files, i.e., *removed* and *added*, determine the insertion and deletion stream. Two further streams share files for clean updates that are optional to execute: *reinserted*, which corresponds to unchanged triples that can be reinserted, and *clear*, which prescribes the delete queries that clear all triples for a resource. Although edits often happen in bulk, this information is not present in DBL, i.e., changes come with no timestamps.

Relevant metadata about the DBPedia Live stream corresponds to license, format, and human-readable description. Listing 6.11 presents the *vocals:StreamDescriptor* [4] that contains basic information about the publisher and the license.[8] The DBPedia ontology and other relevant datasets are linked (lines 5–6). Finally, the VoCaLS file indicates an RSP engine accessible for querying using the vsd:publishedBy.

Now that the DBPedia Live stream has been published, its data can be processed in a continuous fashion. As DBPedia Live already provides RDF data, there is no need to apply any conversion mechanism. Listing 6.12 provides an example using RSP-QL, allowing to query all the entities that were added. Only the added entities will be retrieved, since only the stream with the added statements is queried.

```
PREFIX dbl: <http://live.dbpedia.org/changesets/>
REGISTER RSTREAM <addedStream> AS
CONSTRUCT { ?entity ?p ?o }
FROM WINDOW ON dbl:added [RANGE PT30S STEP PT30S]
WHERE {
  WINDOW ?w { ?entity ?p ?o . }
}
```

Listing 6.12 RSP-QL query sharing DBPedia live added streams

Example 6.1 (Counting New TV Shows) DBPedia Live allows to make various interesting analyses. In this example, the number of new television shows that were added to DBPedia in the last 24hours is counted. As DBPedia uses an underlying ontology schema that models television shows as <http://dbpedia.org/ontology/TelevisionShow>, the RSP-QL query simply needs to fetch all entities that have *TelevisionShow* as concept.

```
PREFIX dbl: <http://live.dbpedia.org/changesets/>
REGISTER RSTREAM <addedStream> AS
SELECT COUNT(?tvshow) AS ?numTvShows
FROM WINDOW ON dbl:added [RANGE PT24H STEP PT24H]
WHERE {
  WINDOW ?w { ?tvshow a <http://dbpedia.org/ontology/TelevisionShow>. }
} GROUP BY ?tvshow
```

Listing 6.13 Counting all TV shows on DBPedia Live added streams

---

[8] http://linkeddata.stream/resource/dbl.

> **Counting the Most Popular Updates**

The analysis of retrieving and counting the types (classes) of most popular updates
in the live stream is left as an exercise. The solution can be found on GitHub page.[9]

6.3.2   Wikimedia EventStreams

```
1  meta: [...], timestamp: 1554284688, id: 937929642, bot: false,
2  type: 'edit', title: 'Q31218558',
3  user: 'Tagishsimon', wiki: 'wikidatawiki'
4  comment: '...', parsedcomment: [...],
5  minor: false, namespace: 0, patrolled: true,
6  length: { new: 5530, old: 5445 },
7  revision: { new: 901332361, old: 756468340 },
8  server_name: 'www.wikidata.org',
9  server_script_path: '/w', server_url: 'https://www.wikidata.org',
```

Listing 6.14   Wikimedia EventStream recentchanges example data

Wikimedia EventStream (WES) is a Web service created at Wikimedia Foundation, i.e., an American nonprofit organization that hosts, among the other,
open-knowledge projects like Wikipedia. In particular, Wikimedia invests financial
and technical resources for the maintenance of projects that foster free and open
knowledge. WES was originally used for internal data analysis and was open-
sourced in 2018. It exposes streams of structured data using SSE, which makes
it an ideal candidate to be processed using the techniques presented in this book.

WES data are gathered from the internal *Kafka* cluster and includes logs and
change-data captures. In practice, WES refers to eight distinct *JSON* streams:
(i) recentchanges (ii) revision-create (iii) page-create (iv) page-property-change
(v) page-links-change (vi) page-move (vii) page-delete (viii) page-undelete. In this
section, the focus will be on the *recentchanges* stream, which models the most
complex content.

Listing 6.14 shows an example of a *recentchange* data item. In WES's recent
changes, the event title links to the Wikidata entity, e.g., in Listing 6.14 the title
points to https://www.wikidata.org/entity/Q31218558.ttl. Each item is timestamped
(line 1) and typed (line 2), helping the introduction of the event concept. Five
kinds of events are possible: "*edit*" (cf. Listing 6.14) for existing page modification;
"*new*," for new page creation; "*log*" for log action; "*external*" for external changes;
and "*categorize*" for category membership change.

Relevant metadata about the WES recentchange include the license, the JSON
schemas, and a human-readable description for the API. Listing 6.15 shows the

---

[9] https://github.com/StreamingLinkedData/Book.

vocals:StreamDescriptor for the *recentchanges* stream. The Wikimedia *Terms of Use* is included as license.

```
wes:recentchange a vocals:StreamDescriptor ;
  dcat:dataset :recentchange ;
  dcat:title "Wikimedia Recentchanges Event Stream"^^xsd:string ;
  dcat:publisher <http://www.linkeddata.stream/about> ;
  dcat:license <https://foundation.wikimedia.org/wiki/Terms_of_Use/en> .
  :recentchange a vocals:Stream ;
     vocals:hasEndpoint :wesendpoint .
```

Listing 6.15 Wikimedia EventStream recentchanges sGraph. Prefixes omitted

In WES, streams are provisioned as JSON data, requiring the setup of a conversion pipeline to foster interoperability. The JSON content can be converted using a mapping language. Listing 6.16 shows an example of RML mapping with a JSON source that is used for the conversion. Line 7 uses `rr:graphMap` to name the RDF graph containing all the triples, which is uniquely done using the event timestamp. Line 10 adds the event type using `rdf:type` predicate and the "type" JSON field as subject.

```
<source> rml:referenceFormulation ql:JSONPath ;
  rml:iterator "$" ;
  rml:source [ a vocals:Stream ; carml:streamName "WikimediaChanges" ] .

<WMM> a rr:TriplesMap ;
  rml:logicalSource <source> ;
  rr:subjectMap [ rr:template "http://www.wikimedia.org/es/{id}" ;
  rr:graphMap [ rr:template  "http://wiki.time.com/{timestamp}" ] ] ;
  rr:predicateObjectMap [
    rr:predicate rdf:type ;
    rr:objectMap [ rr:template "http://....org/es/voc/{type}"] ] [...] .
```

Listing 6.16 RML Mapping for Wikimedia EventStream mapping (Subset)

An extract of result of the mapping is presented in Listing 6.17.

```
<http://www.wikimedia.com/937929642> <http://vocab.org/transit/terms/user> "...";
  <http://vocab.org/transit/terms/server_url> "www.wikidata.org";
  <http://vocab.org/transit/terms/title> <https://www.wikidata.org/wiki/Q312185589
     >;
  <http://vocab.org/transit/terms/comment> "...";
  ...
  <http://vocab.org/transit/terms/wiki> "wikidatawiki".
```

Listing 6.17 RDF result of the RML mapping forWikimedia EventStream

Now that data can be published in RDF, it is ready to be consumed and processed in order to provide some insights.

Example 6.2 (Calculating the WES Stream Rate) A recent challenge triggered a real-time data analysis of WES data, most of which focus on the data visualization aspect.[10] The projects empower simple statistical analyses such as comparison of *bot vs. human* editors, *minor vs. major* changes detection, or categorizing the type of events. This example details a simple yet powerful analysis that computes the rate of the stream. Listing 6.18 shows the RSP-QL query that computes the stream rate of the WES data. Line 2 describes that all the results are counted and divided by 60, which is the size of the window, resulting in the rate of the stream.

```
REGISTER RSTREAM <outputstream> AS
SELECT (COUNT{*}/60) ?ratesec
FROM NAMED WINDOW <win> ON <http://wikimedia.org/recentchanges/rdf>  [RANGE PT60S
    PT60S]
WHERE { WINDOW <win> { ?s ?p ?o } }
```

Listing 6.18   An example of RSP-QL query to compute the stream rate

Example 6.3 (Computing the Most Frequent Users) As a next analysis, the number of changes that each user is conducting is computed. Listing 6.19 shows the RSP-QL query that computes the changes that each user has applied. The query first filters out all the user names and then synthesizes the results by counting the events for each user. The latter is achieved by the group by clause on line 8 and the COUNT built-in on line 2.

```
REGISTER RSTREAM <outputstream> AS
SELECT  ?user (COUNT(?event) AS ?count)
FROM NAMED WINDOW <w> ON <http://example.org/test/rdf>  [RANGE PT10S STEP PT1S]
WHERE
  { WINDOW <w>
       { ?event  <http://vocab.org/transit/terms/user>  ?user }
  }
GROUP BY ?user
```

Listing 6.19   An RSP-QL query that computes the changes applied by each user

⊃   Computing the Most Popular Wikis

Wikimedia contains a number of projects, among others, Wikimedia, Commons Wikipedia, Wiktionary, Wikinews, etc. As an exercise, retrieve a list with the most popular Wikis, in function of the number of changes that were conducted to each wiki project. Make sure to only report the wiki projects that were updated more than twice in the last minute and present them in a descending order, i.e., the wiki with the most changes first. The solution can be found on the GitHub page.[11]

---

[10] http://bit.ly/2FS0mDE.

[11] https://github.com/StreamingLinkedData/Book.

6.3.3   Global Database of Events, Language, and Tone (GDELT)

The GDELT is the largest open-access *spatiotemporal* archive for human society. Its Global Knowledge Graph spans more than 215 years and connects people, organizations, and locations worldwide. GDELT captures themes, images, and emotions into a single holistic global network. GDELT data come from a multitude of news sources using Natural Language Processing techniques.

```
:events a vocals:StreamDescriptor ;
  dcat:title "GDELT Event Stream"^^xsd:string ;
  dcat:publisher <http://www.linkeddata.stream/about> ;
  dcat:description "GDELT Events Stream"^^xsd:string ;
  dcat:license <https://creativecommons.org/licenses/by-nc/4.0/> .
  dcat:dataset :eventstream .
:eventstream a vocals:Stream ;
  vocals:windowType vocals:logicalTumbling ;
  vocals:windowSize "PT15M"^^xsd:duration ;
  vocals:hasEndpoint :eventEndpoint , :oldEndpoint .
```

Listing 6.20 A GDELT stream description using VoCaLS. Prefixes omitted

GDELT consists of three different streams, i.e., *Events*, *Mentions* ,[12] and *Global Knowledge Graphs (GKG)*,[13] each delivered as compressed TSV every 15 minutes:

- The *Event stream* shares geopolitical events appearing in the news. Each event refers to two actors, e.g., nations or public figures, participating in each event, and the action they perform, e.g., a *diplomatic visit*.
- The *Mention stream* records every mention of a particular event in the news over time, along with the article's timestamp. The mention stream is linked to the Event stream by the Global Event ID field.
- The *GKG stream* connects people, organizations, locations, themes, news sources, and events across the planet into a massive network. GKG stream is rich and multidimensional.

GDELT streams use the dyadic format for Conflict and Mediation Event Observations (CAMEO) [2], which contain Global Content Analysis Measures (GCAM). Table 6.1 shows a subset of fields for the Event stream, i.e., GLOBALEVENTID and DATEADDED, that identify the event uniquely and over time; Actors and Event code that link the event to CAMEO; and as an example of GCAM dimension the AvgTone which captures the sentiment analysis of the news article.

GDELT content comes in compressed TSV. Therefore, similarly to the process conducted for WES, a conversion mechanism is necessary. The data conversion follows the RML approach, where CSV data are converted using Mappings.

---

[12] Link to GDELT-Event_Codebook-V2.0.pdf.

[13] Link to GDELT-Global_Knowledge_Graph_Codebook-V2.1.pdf.

Table 6.1 Example of GDELT event data

GLOBALEVENTID	...	Actor1Code	...	EventCode	...	AvgTone	...	DATEADDED
35209457		GOV		020		−3.4188		20190401203000
835209458		LEG		120		1.39860		20190401203000
835209459		USA		040		1.39860		20190401203000

Listing 6.21 shows a sample of the GDELT event mapping. The rr:class is used to assign the gdelt:Event type to the data at line 4. Moreover, the CAMEO is also assigned type using rdf:type at line 7. To model the actors and its type hierarchy, a separate triple map is included at line 12. CARML allows us to extrapolate multidimensional entries of the GKG stream like Themes or GCAM dimensions using custom functions. Additional functions are incorporated that can split the entry content and treat it accordingly. Moreover, the Persons entry has been enriched by retrieving relevant DBPedia entities using DBPedia Spotlight and GeoNames APIs for locations.

```
<GEM> a rr:TriplesMap ; rml:logicalSource <source> ;
  rr:subjectMap [
    rr:template "http://linkedata.stream/resource/gdelt/{GLOBALEVENTID}";
    rr:class gdelt:Event;
    rr:graphMap
      [ rr:template"http://linkedata.stream/resource/time/{DATEADDED}"]] ;
  rr:predicateObjectMap [ rr:predicate cameo:type;
    rr:objectMap
  [rr:template "http://linkedata.stream/ontologies/cameo/{EventCode}"; ]];
  rr:predicateObjectMap [
    rr:predicate cameo:actor;
    rr:objectMap [ rr:parentTriplesMap <Atr1TM> ]]; [...] .
```

Listing 6.21 RML mapping for GDELT event stream conversion (Subset)

Now that the GDELT stream is ready for publication, the data can be analyzed through a number of examples.

Example 6.4 Counting the Events Between China and the USA

Several studies have been running using GDELT. In this example, the number of news events between China and the USA is investigated. This requires the analysis of the GDELT *event* stream and filters out the news event containing both China and the USA as actors. Before *synthesizing* the stream through means of aggregations, the stream is *enriched* by joining with the static data that models the description of the event types using the CAMEO model. Listing 6.22 shows the RSP-QL query that counts the number of events between China and the USA while reporting the type of events.

```
REGISTER RSTREAM AS <TopEventOfTypeT>
SELECT ?eventType ?eventCodeDescription (count(distinct ?event) as ?eventsCount)
FROM NAMED WINDOW <w> ON :GDELTEventRDFStream [RANGE 5PTM STEP5PTM]
FROM NAMED <http://www.geldt.org/cameo>
WHERE{
     WINDOW <w> { ?event   a gdelt:Event;
             gdelt:hasEventCode ?eventType;
             gdelt:hasActor1 <http://geldt.org/cameo/USA> ;
             gdelt:hasActor2 <http://geldt.org/cameo/CHN> .

     GRAPH <http://www.geldt.org/cameo> {
          ?eventType gdelt:hasDescription ?eventCodeDesc .}
}
GROUP BY ?eventCode
ORDER BY ?eventsCount
```

Listing 6.22 RSP-QL enriching GDELT streams. Perfixes omitted

> **Retrieving All Disasters with Large Number of Casualties**

As an exercise, retrieve all events from the GKG stream detailing disasters with a
large number of casualties. The solution can be found on the GitHub page.[14]

6.4   Chapter Summary

In this chapter, we looked at a number of practical examples and exercises that can
be solved using RSP. In particular, we looked at a COVID-19 tracing application
that needs to combine data streams originating from various sources. We also
investigated three real-world data streams, i.e. DBPedia Live, Wikimedia Event
Streams, and the GDELT project. For each of these streams, we showed how the
data can be annotated, published, and processed with various example queries in
RSP-QL.

References

1. Dell'Aglio, Daniele, Emanuele Della Valle, Jean-Paul Calbimonte, and Óscar Corcho. 2014.
   RSP-QL semantics: A unifying query model to explain heterogeneity of RDF stream processing
   systems. *International Journal on Semantic Web and Information Systems (IJSWIS)* 10(4): 17–
   44.

---

[14] https://github.com/StreamingLinkedData/Book.

2. Gerner, Deborah J., Philip A. Schrodt, Omür Yilmaz, and Rajaa Abu-Jabr. 2002. Conflict and mediation event observations (CAMEO): A new event data framework for the analysis of foreign policy interactions. *International Studies Association, New Orleans*.
3. Morsey, Mohamed, Jens Lehmann, Sören Auer, Claus Stadler, and Sebastian Hellmann. 2012. Dbpedia and the live extraction of structured data from wikipedia. *Program* 46: 157–181.
4. Tommasini, Riccardo, Yehia Abo Sedira, Daniele Dell'Aglio, Marco Balduini, Muhammad Intizar Ali, Danh Le Phuoc, Emanuele Della Valle, and Jean-Paul Calbimonte. 2018. Vocals: Vocabulary and catalog of linked streams. In *International Semantic Web Conference (2)*. Vol. 11137. *Lecture Notes in Computer Science*, 256–272. Berlin: Springer.
5. Tommasini, Riccardo, Pieter Bonte, Femke Ongenae, and Emanuele Della Valle. 2021. RSP4J: An API for RDF stream processing. In *The Semantic Web - 18th International Conference, ESWC 2021, Virtual Event, June 6–10, 2021, Proceedings*, ed. Ruben Verborgh, Katja Hose, Heiko Paulheim, Pierre-Antoine Champin, Maria Maleshkova, Óscar Corcho, Petar Ristoski, and Mehwish Alam. Vol. 12731. *Lecture Notes in Computer Science*, 565–581. Berlin: Springer.

Printed in the United States
by Baker & Taylor Publisher Services